2014–2015 IAI 设计奖年鉴

2014—2015 IAI DESIGN AWARD
YEAR BOOK

何昌成 编

华中科技大学出版社
http://www.hustp.com
中国·武汉

CONTENTS 目录

- **4** Preface .... 前言
- **6** APDF Brief Introduction .... APDF 简介
- **7** IAI Award Brief Introduction .... IAI 奖项设置
- **10** The Eighth IAI Design Award Ceremony .... 第八届 IAI 设计奖颁奖典礼
- **12** IAI Design Award Jury Committee .... IAI 设计奖评审委员会

- **16** Interior Space .... 室内空间
- **18** Commercial Space .... 商业空间
- **34** Hotel and Restaurant Space .... 酒店、餐饮空间
- **66** Entertainment and Clubs Space .... 娱乐、会所空间
- **94** Culture and Exhibition Space .... 文化、展示空间
- **120** Office Space .... 办公空间
- **156** Villa Space .... 别墅、豪宅空间
- **178** Apartment Space .... 公寓空间
- **238** Model House Space .... 样板房空间

- **246** Building Space .... 建筑空间
- **282** Industrial Product .... 工业产品
- **296** Projects and Others .... 方案和其他

Perface 前言

何昌成（中国）
Oskar He
(He Changcheng)(China)

APDF 创会主席，现任 APDF 秘书长兼执行委员会主席；
IAI 设计奖发起人、文化部"首届中国设计大展"学术委员会学术委员；
奥地利萨尔斯堡国际夏日美术学院访问学者；
2013 意大利"米兰设计周 -IAI 中国十大设计精英作品展"策展人。

Founding President of Asia Pacific Designers Federation ,APDF Secretary General and President of the Executive Committee of APDF;
Initiator of IAI Design Award ,Member of "China's First Design Exhibition" Academic Committee of the Ministry of Culture；
Visiting Scholar of Austria Salzburg International Summer Academy of Fine Arts；
Curator of 2013 Milan Design Week-IAI China Top Ten Design Elite Exhibition in Italy.

2014 年，第八届 IAI 设计奖颁奖典礼在北京鸟巢文化中心举行，这是一届成功的颁奖典礼，吸引了众多中外媒体的关注和踊跃报导，并被权威电视媒体 CCTV 赞誉为"设计奥斯卡"，近百家媒体更是争相转载报道。

本届 IAI 设计奖竞赛吸引了全球四大洲近二十个国家的设计师和设计机构参赛，其中包括中国（大陆地区和香港、澳门、台湾地区），新加坡、马来西亚，意大利、澳大利亚、瑞典、荷兰、西班牙、英国、墨西哥、厄瓜多尔、日本、斯洛文尼亚等国家。2015 年 4 月 29 日，第八届 IAI 设计奖颁奖盛典在北京鸟巢文化中心隆重举行，由全球设计界具有影响力的设计大师、专家、教授组成的国际评审团经过 3 轮近乎严苛的评选，从近千件参赛作品中精选出 174 件入围作品，最终定出 27 件大奖作品，日本建筑师竹口健太郎的建筑参赛作品《高野山宾馆》摘得了代表本届 IAI 设计奖的最高奖项"IAI 最佳创意大奖"。

本年鉴收录了本届 IAI 设计奖竞赛 117 件获奖作品，作品内容涵盖室内空间、建筑、工业产品三大类别。因为本届竞赛首次将工业产品纳入评审内容，所以参赛数量较少，仅有一件作品获得大奖，占据获奖榜单的获奖作品大部分来自室内和建筑类别，这些获奖作品无疑不是可以成为这个时代创新与创意的最佳经典案例。从本届获奖作品可以看出，就作品的整体水准而言，目前仍以日本、欧美、港台和新加坡等地设计师的作品略显突出，但中国大陆的设计师也成长迅速，尤其是在室内设计领域，进步尤为明显，特别是在空间创意、灯光、材料的应用等方面都达到了相当成熟的水平，接近目前国际最高设计水准。与此同时，许多 IAI 设计奖竞赛获奖者同时也参加了世界其他知名设计大奖，如红点奖、IF 大奖，并获得名次，这都充分证明了 IAI 设计奖与全球顶尖设计大奖在参赛和评审规则已经处在同一个标准上，这是十分令人感到欣慰的，期待通过 IAI 设计奖组织机构和参赛者的共同努力，将 IAI 设计奖打造成为实现参赛者梦想和展示设计智慧和实现自我价值的最佳舞台。当然，不可否认，IAI 设计奖还有许多需要学习和不断成长的地方，何昌成希望随着时间的推移，经验的积累，IAI 设计奖能够不断获得提升和进步，以实现其最终目标。

本年鉴在编辑过程中得到了本届评委会各位专家的大力支持，他们不仅为本届竞赛评出了作品，还在百忙中做了精辟的点评，更要感谢所有的参赛者的热情参与，正是因为他们的参赛作品才有了今天的这本精美的作品集，最后，还要特别感谢本书的出版合作伙伴——华中科技大学出版社，正是因为他们的眼界和真诚合作才共同让这本令人期待的年鉴如期面世。当然，还要感谢所有支持和关注 IAI 设计奖的社会各界人士，因为你们的存在而让这个奖项变得更有价值！

In 2014,the 8th IAI Design Award ceremony was successfully held at Beijing Birds Nest Cultural Center,which is one of the most successful award ceremony,attracting attention and coverage by numerous domestic and international media outlets. It has been regarded as the "Oscar of Design" by CCTV.

This year's IAI Design Award Competition has attracted the participation from nearly 20 countries and design institutions from four continents, including China(Mainland China, Taiwan, Hong Kong, Macao), Singapore, Malaysia, Italy, Australia, Sweden, Netherlands, Spain, UK, Mexico, Slovenia, Japan, and more.The 8th IAI Design Award Ceremony was held at the Bird's Nest Culture Center in Beijing on April 29th,2015.After three rounds of strict assessments,our international jury consisted of the most influential design masters,experts and professors carefully selected 27 top awards from 174 entries. Japan architect, Koyasan Guest House, won the IAI Best Creative Award.

2014 IAI Design Award Winning Works Year Book has included 117 winning entries,which covers interior, architectural and industrial product design, with interior and architectural entries as the majority. This year's competition includes industrial product design for the first time. There was only one industrial product winning the top award,therefore most of the winning works are from interior and architecture category.From the overall level of all the entries, designers from Japan, Europe, America, Hong Kong, Taiwan, and Singapore showed their extraordinary strength. That being said domestic designers are also evolving rapidly, especially in the interior design category where their use of creative space, lighting, material usage, nearly reaches the same level as overseas excellent works. At the same time,many IAI Winners also take part in other design award competitions such as red dot and IF,which fully proves that the judging standard and level of IAI Design Award is comparable to these global top design award competitions,which is very inspiring.To make IAI Design Award the best stage for participants to dream and showcase their design wisdom lot of space for IAI to learn and grow. He Changchen hope that with time and experience goes, IAI can make constant progress and improvements, thus achieving its final goal.

This year book is supported by this year's judges, who not only selected the best entry but also gave sharp comments. Thank you to participants whose wonderful projects made the year book possible. A very special thank you to the publisher-Huazhong University of Science and Technology Press house,whose insight and sincere cooperation made this year book come out on time. For all friends who supported IAI Design Award,your existence has made the award more valuable.

APDF Brief Introduction APDF 简介

APDF
亚太设计师联盟
ASIA PACIFIC DESIGNERS FEDERATION

亚太设计师联盟（APDF）简介

筹建于2006年，正式成立于2008年，2009年经由香港政府警署社团中心核准并正式更名为亚太设计师联盟（APDF），是一个民间非营利组织，其核心宗旨为促进设计相关产业协同发展，通过联盟行动举措解决所在国家和地区社会经济发展和需求问题，提升人民生活品质。APDF由来自世界各国设计领域知名设计大师、专家、学者、教授及各国设计精英组成，是一个在全球范围内影响力日渐上升的国际专业设计学术团体组织。

愿景与使命

促进设计与相关产业协同发展，通过联盟行动举措解决所在国家和地区社会经济发展和需求问题，提升人民生活品质。针对不同国家和地区的政治、社会和文化差异，为亚太地区设计发展提供多样性服务。以一个学术团体组织和非党派及非政府组织发出联盟的设计声音。倡导将设计作为促进人类与环境和谐共处、实现最佳利益的有效途径。APDF以一个新的国际性设计团体组织身份定位自身的发展及终极目标，倡导并逐步建立符合未来社会及行业发展的"设计标准"、"成就认定"、"知识产权保护"、"可持续发展"、"传统工艺与文化"与"高新技术"。引导和培养公众和个人将设计从业者及客户视为社会上一个重要和有价值的职业意识。提倡专业设计师以负责任的方式提供设计服务和品质保障。发现并提携优秀设计作品，定义最佳设计案例与管理实践。通过学术引领和行动举措，促进将设计作为改变社会经济发展和生活方式的一种重要力量和有效途径；在世界各国开展高质量的设计教育与实践活动，促进设计教育与职业发展，分享最佳实践经验，提升学生设计能力。建立与世界其他国际组织和相关设计机构的合作，促进全球设计行业的发展。

学术成果

由APDF创会主席何昌成先生于2006年创立的IAI设计奖竞赛作为APDF重要创意活动已经成功举办8届，被中外媒体赞誉为"设计界的奥斯卡"，是亚太地区乃至国际具有影响力和美誉度的设计竞赛之一。
APDF先后成功举办了8届IAI设计奖竞赛，发现并提携了一大批国内外设计新秀和优秀作品。先后在国内外组织了数十场中外学术论坛和形式多样的设计交流活动。多次组织中国设计师赴奥地利、德国、迪拜等国家和地区进行交流考察和研修学习，促进了中国设计文化与世界设计文化的融合与发展。

Brief Introduction

Asia Pacific Designers Federation(APDF) was built in 2006 and was officially founded in 2008.It is a non-governmental and non-profit organization aiming to promote the development of design related industries and find solutions to design related needs of the world. The company was formally approved by the community center of Hong Kong Government and its name was settled in 2009. Consisting of master designers, specialists, scholars, and professors from around the world, APDF hopes to lift the living standard of our communities. It has become an international professional academic organization with an increasing influence on a global scale.

Vision, Mission, and Values

Promote coordinated development ofdesign and related industries and devoted to drive development of national and regional economy; Provide diverse services for design development in the Asia Pacific region according to politic,social,and cultural differences of different countries and regions; Advocating design as an effective method which can promote harmonious existence of humanity and their environment;Aim to be an academic, non-party, and non-governmental Federation;APDF, a renewed international design organization, advocates and builds design standards, achievement certification,intellectual property right protection, sustainable development, as well as respects both traditional craft & culture and innovative & high technology in accordance with the development of society and industry in the future; Guiding and fostering professional consciousness, improving level of design and professional integrity, as well as regarding designers as a promising vocation; Calling for professional designers with a responsible attitude towards design services and quality assurance; Looking for and promoting excellent design works, defining the best design cases and best practice of management; Under the guidance of academic leadership and actions, promoting design as a significant role and effective way in changing social and economic development and lifestyles.Conducting high quality design education courses and practical activities all over the world, enhancing development of design education and career, sharing best practice and improving ability of design; Cooperating with other international organizations and related design institutions from all over the word, enhancing global design communication and cooperation;

Academic Achievements

IAI Design Award was founded in 2006 by APDF founding president, Oskar He.Since its inaugural year, the event has been held successfully in the following eight years. In that time span the event has recognized and developed the works and careers of hundreds of designers both young and world renowned. The event has been recognized as "the Oscars of Design" by both domestic and foreign media outlets and now IAI Design Award has entered the global design stage.
Inaddition to the work completed with IAI Design Award, APDF has also promotedthe integration and development of Chinese design culture and world design culture by conducting many academic investigation, training activities, and design forums throughout in China, Austria, Germany, the United Arab Emirates,and other countries around the world. Such events and activities have allowed for the continued communication and development of cultural and design related cooperation.

IAI Award Brief Introduction IAI 奖项设置

奖项设置

IAI 最佳创意大奖
IAI 年度最佳设计机构
IAI 杰出设计大奖
IAI 最佳设计大奖
IAI 设计优胜奖
IAI 设计之星奖

Awards Setting

IAI Best Creative Award
IAI Annual Best Design Agency Award
IAI Best Design Award
IAI Outstanding Design Award
IAI Design Excellence Award
IAI Design Star

新增奖项设置

IAI 评审团特别奖
IAI 最佳概念创意奖
IAI 最佳材料创意奖
IAI 最佳人文关怀奖
IAI 最佳环境友好奖

New awards Setting

IAI Special Jury Award
IAI Best Concept Creative Award
IAI Best Material Creative Award
IAI Best Humanistic Care
IAI Best Environment Award

IAI 设计奖（IAI Design Award），简称"IAI"，是亚太地区乃至国际具有影响力和美誉度的设计大奖之一，该竞赛以独树一帜的参赛内容和奖项设置并紧贴时代主题而日益受到全球设计界的关注。IAI 设计奖始创于 2006 年，先后成功举办 8 届设计竞赛，现已被视为拥有高识别度和含金量的国际设计大奖。

该赛事自 2014 年起升华为全域设计的 IAI 设计奖，其三个字母（IAI）完整传达了大赛的内容，即室内、建筑、工业产品，并延展至传播设计、产品概念设计等多学科、跨专业设计领域。主办方希望通过举办 IAI 设计奖竞赛挖掘社会各领域的创新设计，并进行广泛地宣传推广，努力提高设计的质量，使之被社会认可和接受，积极寻找设计作为引导未来力量的价值。

IAI 设计奖将致力于寻找和发现符合当下生活方式的优秀设计作品，同时为获奖作品和获奖者进行品牌宣传和市场推广，将为设计概念转化为商品，为获奖创意产品和商业化合作牵线搭桥，IAI 设计奖奖项分别由代表 IAI 最高荣誉和学术高度的 IAI 创意奖、代表时代创新和商业价值的 IAI 设计奖组成，奖项内容涉及室内设计、建筑设计、工业产品设计、传播设计、产品概念设计五大部分，并含有设计论坛、设计出版、设计贸易洽谈会、设计展示、设计推广和顾问服务等内容。

IAI Design Award is the most influential and reputable design award in Asia Pacific area and on the global stage.The competition with its unique award contents and setting has getting more and more attention from global design circles.IAI Design Award established in 2006 ,has been held successfully for eight years.It has been viewed as an international design award with a high degree of recognition and value.

In 2014, IAI Design Award was wholly reformed into a dynamic design competition encompassing five professional fields interior design,architectural design and industrial product design, as well as communication design and product concept design. The host hopes to excavate social innovative ideas in the fields of design, promote the reward widely, take efforts to improve the value of design and make its foothold in society, actively seeking the value of design as a guider in the future power, which aims to advocate designers to uphold social responsibilities.

IAI Design Award devotes to find excellent works at the current, promoting and marketing for winners and their winning projects. It will be committed to transferring the winning design concept into a commodity, linking the award-winning creative concept to commercial cooperation. IAI Design Award includes interior design, architecture design, product design, communication design and product concept design of five major components, also providing with design forum, design publishing, design and trade fair and exhibition, design promotion and advisory service, etc.

第八届 IAI 设计奖颁奖典礼
The Eighth IAI Design Award Ceremony

第八届 IAI 设计奖颁奖典礼
The Eighth IAI Design Award Ceremony

第八届 IAI 设计奖颁奖典礼
The Eighth IAI Design Award Ceremony

第八届 IAI 设计奖作品展览
The Eighth IAI Design Award Winning Entries Tour Exhibition

IAI Design Award Jury Committee IAI 设计奖评审委员会

Since establishment of Asia Pacific Designers Federation (APDF), on this academic platform connected international designers , APDF has organized successfully eight design competitions in non-commercial mode and promotes the communication of design vigorously and transnationally. Through these academic activities closed to theme of the time, which can unite members and make new connection, not only APDF grows constantly, but also his social influence further strengthens. Therefore IAI Design Award which is held by APDF, beyond the scope of the Asia-Pacific region, becomes an important events which obtains attention of designers all over the world .

There are a number of designers participating in 2014 IAI Design Award.The organizer also invited the most influential designer masters,experts and scholars consisted of the jury panelv. Because of the strict selections, the results of the competition are of high authority and orientation. The 2014–2015 IAI Design Award Year Book is not only a summarize of of all the entries but also helps to advertise and promote the excellent works in an effort to make innovative design as a force to push social and economic development, which make designers become the practitioners to achieve sustainable development goal.Our life will become more happy and beautiful because of the constant innovative design.

IAI 亚太设计师联盟自成立以来，在这个连接各国设计师的国际学术平台上，以非商业化的运作模式成功举办了8届设计竞赛活动，有力地推动着跨越国界的设计文化交流。通过这些紧贴时代主题的学术活动，凝聚人气，扩展人脉，使 IAI 组织规模不断壮大，社会影响力迅速提升，走出亚太地区成为世界范围设计师广泛关注的重要组织。

2014 年 IAI 设计奖的参赛者众多，组委会邀请世界具有影响力的设计大师、专家学者组成评审团，经过严格的评审、选拔，使这次大赛的评审结果具有权威性和导向性。在此基础上出版的《2014—2015 IAI 设计奖年鉴》，既是总结汇报，又是对评选出的优秀设计作品的宣传和推广。IAI 组织努力要让创新设计真正成为推动社会进步和经济发展的动力。让设计师真正成为实现可持续发展目标的践行者，让生活因不断涌现的创新设计而幸福美好！

Kees Spanjers think as designers, they are not designing to win awards. They are designing to enhance people's lives and to bring beauty, comfort and sustainability to the environment. But from time to time it is good to judge each other as professionals, and look critically at each other's works. That's what they have been doing as international judges in this years IAI Awards, and they are amazed by the quality of the works they have seen. They encourage all the participating designers to go on with the good work that you do here in Asia,Kees Spanjers thinks the work is outstanding

我认为，作为设计师，我们的追求不是获奖，而是为了提高人们的生活质量，并让环境变得美丽、舒适、永恒。但有时，以专业的身份来互相评价，或者评论彼此的作品，也是很有意义的。这就是作为 IAI 奖国际评委所做的事情。参赛作品的高质量令评委们十分惊喜。评委们鼓励所有的参赛设计师继续他们的设计道路，因为他们的作品是很优秀的。

Design is the process of making thing better for people more useful,usable and desirable. and act of helping people to do something. Design is respect to the other people, and designer is respect for the environment of the globe.

设计是一个创造对人类更有用、更适用、更合意的事物的过程，同时也是一个帮助人们完成某些事情的过程。设计必须尊重其他人，设计师必须对地球环境怀抱崇敬之心。

张绮曼（中国）
Zhang Qiman（China）

亚太设计师联盟（APDF）名誉理事长；
中国室内设计泰斗；
中国环境艺术设计学科创建人；
中央美术学院建筑学院教授、博士生导师；
清华大学美术学院博士生导师。

Honorary President of Asia Pacific and academic leader of Chinese Environment Art Design;
Doctoral supervisor of Environment Art Design;
The doctoral supervisor and professor of Central Academy of Fine Arts in the School of Architecture;
The doctoral supervisor of the School of Fine Arts of Tsinghua University and National Academy of Arts.

凯斯·思班尼斯（荷兰）
Kees Spanjers
（Netherlands）

亚太设计师联盟（APDF）联合主席；
荷兰国家室内设计协会主席、欧洲室内设计师联盟（ECIA）主席（2004—2008 年）

2004–2009, Co-chairman of Asia Pacific Designers Federation (APDF);
President of National Institute of Interior Design;
Chairman of the European Council of Interior Architects from 2004 to 2008.

李淳寅（韩国）
Soon-In Lee
（South Korea）

亚太设计师联盟（APDF）联合主席；
1990 年都柏林欧洲设计中心主席；
韩国设计振兴院（KIDP）本部长；
韩国弘益大学国际设计学院的院长及研究生学院设计管理专业的讲师。

Co-Chairman of APDF;
1990,Serving as president of the LG Europe Design Center in Dublin. Minister of Korea Institute of Design Promotion(KIDP).
Dean of international design academic, Hong-Ik University;
lecturer of Graduate school of design and management.

IAI Design Award is a proud award in Design World.Not only its judging criterion is the highest international judging criterion,but also it requires winning works to reflect some design concepts:environmental protection, green, low carbon and sustainable, in order to guide the future society to be more civilized, friendly and harmonious.Wang Min thinks that this kind of award is the most valuable, and it is also worthy of praise.Therefore IAI Design Award is the biggest praise and affirmation, not only for works,but also for participants.

IAI 设计奖是一个足以让设计界为之骄傲的设计奖项,不仅因为它始终以国际级别的评审标准来评定作品水准,并要求获奖作品体现生态、环保、绿色、低碳、可持续的设计理念要求,以此引导未来社会共同向更加文明友好、和谐的生活方式转变,王敏认为这样的奖项是最有价值的,也是最值得推崇的。因此获得 IAI 设计奖无论是对作品本身还是参赛者个人都是一种褒奖和肯定。

What makes IAI Design Award unique is that it has grown and expanded from interior design to architectural design and industrial product design, as well as communication design and product conceptual design with competent expert juries and excellent design. This is why IAI Design Award has received more and more attention and recognition from international design field.Liu Guanzhong expects and believe that IAI Design Award will be viewed as an international design competition with a high academic level and recognition.

IAI 设计奖与国内外许多奖项不同之处在于它是一个有着自己鲜明成长轨迹的国际设计大奖,从室内到建筑再到工业产品,并延伸至传播设计和概念设计,一步一个脚印走出了一条独具特色的成长之路。客观、公正的评审和高含金量的获奖作品都是该奖项越来越受到国际设计界关注和认可的理由。期待并相信 IAI 设计奖未来将能够实现成为设计领域最具学术高度和影响力的国际设计大奖的目标。

Zuo Hengfeng is looking forward that IAI can be an important platform that designers can communicate with each other,share information and encourage each other.He hopes IAI can be created as the most authoritative and high quality award like IF and Red dot.

左恒峰期待 IAI 能够成为一个设计师相互沟通,分享信息,以及彼此激励的一个重要的职业化平台。希望 IAI 能够被打造成与德国 IF、红点奖相媲美的在亚洲具有最高权威性和最高质量的重要奖项。

"This year I am a judge for the IAI Awards. We had some good submission and I'm happy with how each year, we have greater submission numbers. It is supported by CAFA which is a fantastic university in Beijing and I'd encourage people to put this competition on their radar. As it's the home of the Birds Nest Stadium.You can't ask for a better stadium for the awards and the Chinese design community is getting better and better each and every year".

今年作为 IAI 的评委,Ryan Roth 看到了一些很优秀的作品,很开心看到每一年都有更加优秀的作品参与进来。这次大赛得到了北京中央美院的支持,他希望鼓励更多的人了解这个比赛。对于 IAI 而言,没有能比在鸟巢颁奖更好的地方了,中国设计团队也变得一年比一年更好。

王敏(美国)
Wang Min(U.S.)

亚太设计师联盟(APDF)总理事长(2015—2017 年);
中央美术学院设计学院院长、教授、博士生导师;
北京奥组委形象与景观设计总监;联合国科教文组织中国委员会评审专家、世界经济论坛(达沃斯)设计创新理事会理事。

Council President of Asia Pacific Designers Federation(APDF) (2015–2017).
Dean,Professor and Doctoral Supervisor of Design College of Central Academy of Fine Arts;
Image and Landscape Design Director of the Organizing Committee of Beijing Olympic,UNESCO Chinese Committee Evaluation Expert,Member of of the World Economy Forum Davos Design Innovation Committee.

柳冠中(中国)
Liu Guanzhong(China)

APDF 副主席和理事会成员;
清华大学美术学院责任教授、博士生导师;
中南大学艺术学院兼职教授和博士生导师;
中国工业设计协会副理事长兼学术和交流委员会主任。

The Vice-President of APDF Council;
The Professor and the Doctoral Supervisor of the College of FIne Arts of Tsinghua University;
The Vice Council President and Dean of Communication Comiittee of China Industrial Design Association;
The professor and doctrial tutor of Academy of Arts ,Central South Universirty.

左恒峰(中国)
Zuo Hengfeng(China)

设计艺术学博士,硕士研究生导师;
清华大学美术学院工业设计系副教授;
英国南安普敦索伦特大学设计学院客座教授。

Art Design Doctor, Academy of Fine Arts;
Associate Professor of the Tsinghua University ,Academy of Fine Arts,Industrial Product Design;
Visiting professor of Southampton Solent University.

瑞恩·罗斯(英国)
Ryan Roth(U.K.)

瑞恩·罗斯现任罗斯管理的首席执行官;
英国伦敦凯斯乔荻设计有限公司全球总经理;
被 2013 年《卫报》和《观察家报》评为全球最值得关注的 500 位文化界最有影响力的人物之一;

Present CEO of Roth Management; the General Manager of UK London Marques and Jordy;
One of the top 500 cultural influencers in the world to watchby The Guardian and The Observer in 2013.

IAI Design Award Jury Committee IAI 设计奖评审委员会

IAI Design Award was wholly reformed into a dynamic and multi-disciplinary design competition from the old IAI simply encompassing interior design, so organizers would like to sincerely invite more design-lovers to participate in IAI Design Award.

今年的 IAI 从过去单纯的室内设计评比，蜕变更多样性的跨界设计领域奖项，所以我们希望能够邀请更多的设计爱好者热烈参与我们的 IAI 设计奖。

The IAI Awards present a wonderful opportunity to showcase your work in The Asia Pacific Region and on the International stage. It represents a variety of multi- disciplinary collaboration across the design industry and promotes the recognition of the value of design. Take the opportunity to support the Awards programme and compliment the exposure of your studio into your region.

IAI 奖提供给你们一个绝好的机会，在亚太地区和国际舞台上来展示你们的作品。IAI 奖倡导整个设计行业的跨领域、多样化合作，提升对设计的价值认可。而且借此机会支持获奖项目，并且赞扬你们的工作室在相关领域的表现。

IAI Design Award has increasingly become an international influence award, and I hope more international designers to attend this activity, making our IAI design can truly represent the Asia region, the Pacific and the design level of the world.

IAI 设计奖已经日益发展成为一个具有国际影响力的一个奖项，我希望有更多的国际设计师来参加这个活动，使我们 IAI 设计奖能够真正代表了亚洲地区、太平洋地区和世界的设计水平。

I am very excited to see this team. These products are different and have quality but I haven't seen many creative products in interior and architectural design categories.

看到这个团队，很兴奋，都有不同的，有概念的团队，质量是有的，但是非常创意的产品，在室内和建筑领域十分具有创意的作品，不是特别多。

陈光雄（中国台湾）
Adam Chen
(Taiwan, China)

IAI 亚太设计师联盟国际理事会副理事长；
台湾圆镜建筑设计事务所总监；
建筑师、规划师、室内设计师。

Vice-Chairman of IAI Asia-Pacific Designers Federation.
President and Director of Econergy +sustainable living design consultancy integration.
A famous Architects, Planners, Interior designer.

安东尼奥·詹诺内（澳大利亚）
Antonio Giannone
(Australia)

Tectvs 的董事；
澳大利亚建筑师；
阿德莱德大学建筑学院建筑环境专业的兼职副教授；

A Director of Tectvs ;
He is a Fellow of the Australian Institute of Architects;
He is currently an Adjunct Associate Professor, School of Architecture and Built Environment at the University of Adelaide.

王庚飞（中国）
Wang Gengfei (China)

空间设计师与独立学者；
"中国民间美术遗产保护与研究中心"学术委员会委员；
《中国民间美术遗产普查集成》总编委；
《世界园林》杂志执行编委；
众多综合设计项目的总设计师。

Space Designer and Independent Sttcholar;
The members of 'China Folk Fine Arts Heritage Protection ;
The general editorial board member of 'Chinese Folk Fine Arts ;Heritage General Survey, ;
The executive editorial board member of 'World Landscape' magazine .
The general designer of numerous design projects.

何新城（荷兰）
Neville Mars
(Netherlands)

何新城建筑事务所是一家中荷合资公司，在孟买和上海都开设了分支机构；
其客户有 BMW、古根海姆博物馆、荷兰皇家壳牌、大众汽车、Walltopia、博洛尼等知名品牌。

MARS Architects is a Sino-Dutch company, have branches in Shanghai and Bombay.
MARS architects' clients include froward looking brands such as the BMW and Guggenheim Museum, Royal Dutch Shell, Volkswagen Group, Walltopia and BOLONI.

Design from all around the world actually is directed towards sustainability. The Jury of IAI Design Award focus on three perspectives of design -profession, prospect and potential.Design is to serve the society,and it is one industry that can bring happiness to the society.

其实全世界的设计都是朝可持续的方向发展的，IAI 的评委可以从三个角度，即从很专业、很未来、很有潜力的角度来评价作品，设计其实是服务社会的，它是可以带给社会欢乐的一个行业。

After watch these works, Chang Zhigang believes they represent the highest level in Asia.The judges here also had a heated debate, which represents a different point of view.It represents the international standard ,so such a review activity is convincing as well.

看到这些作品，确实还是代表了亚洲的最高水平，评委们在这里也进行了激烈的争论，代表了不同的观点，当然也代表了一个国际化的水准，所以这样一个评审活动还是令人信服的。

张清华（中国台湾）
Ching-Hwa Chang
（Taiwan, China）

九典联合建筑师事务所主持人；
曾在成功大学，台湾科技大学等校兼任教授级专家，并为美国绿色建筑协会会员。

The principal architect of Bio-architecture Formosana.
She has taught in National Cheng Kung University, National Taiwan University of Science and Technology and the member of USGBC.

常志刚（中国）
Chang Zhigang（China）

中央美术学院建筑学院副院长、副教授、硕士生导师。

Deputy dean、Associate professor、and Master tutor of Central Academy of Fine Arts School of Architecture。

INTERIOR SPACE
室内空间

Commercial Space	18
Hotel and Restaurant Space	34
Entertainment and Clubs Space	66
Culture and Exhibition Space	94
Office Space	120
Villa Space	156
Apartment Space	178
Model House Space	238

18	商业空间
34	酒店、餐饮空间
66	娱乐、会所空间
94	文化、展览空间
120	办公空间
156	别墅、豪宅空间
178	公寓空间
238	样板房空间

Commercial Space
商业空间

Class - e Convergency in Taichung ,Taiwan	20
Warm Time of Summer Earlier ——Jolyvia Skin Care Salon	22
Xuhui Haishang International Building-selling Center	24
PRIM4 Hair Salon	26
Taiwan Mobile Digital Life	28
LEICHT	30
Jongtay Furniture Expo Center	32

20	台湾 . 台中同学汇
22	初夏的温暖时光——娇莉芙养颜馆
24	旭辉海上国际售楼中心
26	PRIM4 美发沙龙
28	台湾大哥大数字生活
30	筑韵
32	中泰家具博览中心

Commercial Space 商业空间

台湾．台中同学汇
Class – e Convergency in Taichung, Taiwan

项目地点：台湾省台中市西屯区河南路四段171号 /Location : 14F, No.7, Zunxian Street, North District, Taichung, Taiwan
项目面积：720 平方米 /Area : 720 m²
公司名称：九号室内装饰有限公司 /Organization Name : Nine Studio
设 计 师：李东灿、陈俊凯、江若齐 /Designers : David Lee, Jiun-Kai Chen, Jo Chi

IAI 最佳设计大奖

九号室内装修有限公司
Nine Studio

李东灿　David Lee　陈俊凯　Jiun-Kai Chen
江若齐　Jo Chi
中国台湾　Taiwan, China

9 STUDIO 成立于2005年。
以集合住宅、独栋住宅、商业空间、办公室及医疗院所的建筑及室内设计为主，多元经营，合伙人制设计公司。我们认为每个空间应有其自身散发出的一种特有的表情，注重是空间本身使用性质及使用者对于空间的使用态度。

9 STUDIO is established in 2005.
9 STUDIO is a professinal team that specializes in integrative design . They are a diverse partnership company that offers services mainly including the architechtural and company that offers services mainly including the architectual and interior design for collective housing,individual housing,commercial space,office and medical insititution.
They believe each space expresses a unique sensation which is only built upon their perception of the composition of the space,but more importantly upon the use of the space and the user'sattitude towards space use.

本案设计试图打破限制式的学校教育和规格化的教学方式，创造优质、舒适的自习场所和交流环境，让孩子从此拥有学习自主权。接待大厅如雕刻般个性化的接待柜台，连带着高吧台设计。墙面有着大范围的涂鸦墙，可随意彩绘，再搭配有着跳跃感的灯具，镂空墙面的光线，增添了轻松的气息。学生休息之余或站或坐，自在交流，开阔的空间提供了弹性使用的可能。该设计以一道强化清玻璃隔间墙，划分出动与静的两个空间属性。平面图左侧的完整区域，是学生自习区。偌大的阅览室，左右摆放了各式各类课外读物，犹如一间小型书店。除"一"字形的书桌设计可供学习外，刻意安排了沙发区，随手拿起一本书，窝坐于沙发中，也是在学习课业之余的一种减压的享受方式，从而转变了对传统教育的严肃空间观念。

This project is to break the form and standards of traditional school and normalized way of teaching, to create a comfortable self-study field and communication environment, so children have learning empowerment. Reception hall installed with personalized reception bar counter, a wide range of graffiti wall, colored drawing on wall, special planning of lamps and lanterns, the wall pierced to release its interior space with a positive and light filled shadow to bring atmosphere of relax, which satisfies the students' demands to host diverse learning activities such as theme events, musical, dancing, group activity rehearsals, online readings and on-site readings. The field is divided into 2 parts (dynamic and static environment) by one glass wall. The left on the plan is Self-study Area which includes a large reading room with all kinds of reading materials,like a small bookstore."—"shaped desk designed for several students, and students can read on the sofa area, which is a way to relieve pressure on the taking a break from school,and changes the serious concepts of traditional education.

初夏的温暖时光——娇莉芙养颜馆
Warm Time of Summer Earlier——Jolyvia Skin Care Salona

项目地点：广东省广州市 /Location : Guangzhou, Guangdong
项目面积：530 平方米 /Area : 530 m²
公司名称：杨佴环境艺术设计有限公司 /Organization Name : Zhuhai Yang Er Environment Art Design Co., Ltd.
设 计 师：杨佴 /Designer : Yang Er

IAI 设计优胜奖

杨佴 Yang Er
中国 China

杨佴环境艺术设计有限公司创办人、首席设计师；
深圳室内设计师协会常务理事；
从事设计工作 20 年；
2013 年中国国际大学生空间设计大奖"ID+G"
金创意奖特邀专业导师；
2005 年创立杨佴环境艺术设计有限公司；
2003 年进入香港高文安设计公司，期间任职手绘主笔、设计总监；
1995 年先后在北京、珠海创立个人设计公司。

Chief Designer and Founder of Yang Er Environment Art Design Co. Ltd.;
Executive Member of Shenzhen Association of Interior Designers;
2013 Professional Instructors of "ID+G";
2005 Created Yang Er Environment Art Design Co. Ltd.;
2003 Design Director of Hong Kong Kenneth Ko Design Co.,Ltd.;
1995 Created personal design company in Beijing and Zhuhai successively.

依地势而建的改造和半开放的建筑结构，将现代简约的美学风格与越秀山的自然环境以一种时尚的方式进行融合，似"大隐隐于市"的都市隐士。

通过前厅进入露天的中庭花园，自然光透过树木缝隙洒下院内，形成斑驳、迷离的阴影，给繁华的商业都市带来一份难得的幽静。借此为景的是内厅，也是客户休息厅，这里有悬吊的以自然麻石的弧形屏风，丰富了空间的动线，在自然光线影响下的水泥圆形柱子，被赋予了诗意，既质朴又时尚。庭院式护理店，更容易让人与自然环境产生心灵上的沟通，在这里，娇莉芙养颜馆将时间定格在初夏的温暖时光。

According the situation to reform, this half open building structure, it is with Aesthetic style of modern simplicity and contracted with the Yuexiu mountain natural environment become a kind of fashion accommodation, as a city hermit of "Greatest hermit often lies the downtown ".

Enter the open-air atrium garden by the lobby, and suddenly see the light, and the natural light is glimmering into the yard through the air and gap of the leaves, forming a blurred mottled shadow, there is a rare quiet in the bustling commercial city. Here is inside of the hall, is also a customer lounge, there are some natural granite screen curved of be overhung, enriches the movement and exchange of the spatial. In the natural light, the cement circular columns are endowed with poetic, simple and fashion. In the care store of type courtyard, Easier to make a communication between people and the natural environment by spirit, here , Jolyvia Skin Care Salona times frame in the summer time.

Commercial Space 商业空间

旭辉海上国际售楼中心
Xuhui Haishang International Building–selling Center on the Sea

项目地点：上海市杨浦区控江路 2018 号 /Location : No.2018, Kongjiang Road, Yangpu District, Shanghai
项目面积：600 平方米 /Area : 600 m²
公司名称：IADC 涞澳设计公司 /Organization Name : IADC Designers. Ltd.
设 计 师：张成喆 /Designer : Alessio Zhang

IAI 设计优胜奖

张成喆 Alessio Zhang
中国 China

IADC 涞澳设计公司创始人，首席设计师；
任国际建筑协会（ICU）中国区常务理事；
近年正着力打造跨领域设计平台；
作品常发表于海内外专业媒体及出版物上，并著有个人作品集"喆思空间"。

Founder and chief design in IADC design company in Australia;
Standing director in International construction Union (ICU) in China;
Create the interdisciplinary design platform in recent years;
Works often published in professional media and publications at home and abroad, and author of a personal portfolio "Zhesi space".

旭辉是中国驰名的专业地产开发商，始终坚持开发"经典""舒适""健康""绿色"的时代精品。该项目的建筑形体由两个长方体组成，由此形成了两个狭长的空间，如何将这两个空间巧妙地串连起来，形成一致性是本案室内设计的焦点和难点之所在。设计师张成喆以两个长方形空间的内侧转角墙体为主立面。然而，在狭长空间中，实体墙往往给人窒息的感觉。于是，一道由线、面组成的镂空墙面应运而生，形成类似庭园中竹篱墙的观感，疏密有度，间或以植物点缀其中，仿佛置身于户外庭园之中。这种令人意想不到的"篱墙"效果正是本案的点睛之笔，以抽象的表现手法，几何构成关系为元素，营造室内庭园的意象，在某种意义上，与印象派风景画的表现方式不谋而合，故又名为"庭园印象"。

CIFI Group, which is a well known Chinese property developer, always upholds the corporate philosophy of "Classic, Comfortable, Health and Green". The architectural structure of this project is composed by two rectangular spaces. The biggest difficulty and challenge is how to connect these two narrow spaces. Designer Alessio Zhang was inspired by the impressionist art takes the corner of the wall as main building, and transformed into a landscaping wall. However, hypostatic wall easily gave people a feeling of depression. Instead, a hollowed-out wall was emerged to create a feeling of hedgerow. Here, people can feel like they're in the outdoor garden. Better yet, light effect was skillfully placed in the hollowed-out wall to achieve a fascinating visual effect. This design is also the finishing touch to the space. Through abstract expression, Zhang creates a feeling of indoor garden. In the technique of expression, this design coincides with impressionism, so this project also known as "Garden Impression".

Commercial Space 商业空间

PRIM4 美发沙龙
PRIM4 Hair Salon

项目地点：台湾省台北市大安区安东路 /Location : Andong Road, Daan District, Taibei, Taiwan
项目面积：30 平方米 /Area : 30 m²
公司名称：优玛设计 /Organization Name : YOMA Design
设 计 师：许祐承，张美铃 /Designer : Xu Youcheng (Yoshi Xu), Zhang Meiling (Maki Zhang)

IAI 设计优胜奖

优玛设计 YOMA DESIGN
中国台湾 Taiwan, China

YOMA DESIGN 是 YOSHI 和 Maki 所创办的建筑室内设计公司，所秉持的精神就如同 "YOMA= 有吗 " 的谐音，一种幽默的设计方式。因为你们找 YOMA，才有 YOMA 的设计风格产生，YOMA 以不设限的设计风格在创作更多的设计产值，就如同 YOMA 的融合。他们一直以调整和创新的设计方式前进。

YOMA DESIGN is an architecture and interior design company founded by Yoshi and Maki. By holding the spirit as homophonic "YOMA = You Ma", it's a humorous way of design.
The design style of YOMA is based on your demands "YOMA". YOMA pursue unlimited ideas, create with minimalism. It's just like integration of YOMA. They always make progress by adjusting and innovating in the way of design.

PRIM4 美发沙龙坐落于台北市大安区，位于繁荣的商业地带，如何能让一个老社区动起来，这是 YOMA 所面临的第一个问题。PRIM4 美发沙龙希望提供给顾客一个互动有趣的休息场所。YOMA 利用多种形状使设计具有的光线变化，颠覆传统的顶棚，以不同形状的镜面去打破一般人的想法，给人更无限的想象。这个城市虽然冷漠，但发廊却如此生动。同时设计师将童趣带回成年人的生活，让整个空间更有意思。PRIM4 占地面积 30 平方米，却清楚地划分为 5 个不同的主题区域，尤其是设置了 2 个 VIP 室由布帘和线帘的隔绝，保护隐私却不感狭隘。这是 YOMA 期许 PRIM4 能带来的印象性、功能性、趣味性、专业性，是 YOMA 让 PRIM4 美发沙龙变成令人惊叹的发廊设计。

PRIM4 locates at the central of Taipei City, where is full of business activities. How to enliven this old community.It's YOMA's first question. PRIM4 provide their customers a salon which is childlike but experienced. Based on hair designers' specialty, YOMA play the space with various lighting, reversed lawn ceiling, and non-square mirrors to break the rules. The city is indifferent, but the salon is lively. They integrate childhood into adulthood.They make it interesting rather than boring. PRIM 4 is only 30 square meters, but it distinct to five characteristic work spaces. Especially, there are two VIP rooms were isolated masterly by line curtain and drapes, giving privacy but penetrable. This is what YOMA expects PRIM4, functional, interesting, and professional. PRIM4 is a "WOW" hair salon.

YOMA DESIGN
X
PRIMA

台湾大哥大数字生活
Taiwan Mobile Digital Life

项目地点：台湾省台北市信义区烟厂路 80 号 1 楼 /Location : Building 1, No.80, Yanchang Road, Xinyi District, Taibei, Taiwan
项目面积：330 平方米 /Area : 330 m²
公司名称：金玉岑空间设计公司 /Organization Name : K+SPACING
设 计 师：金禹岑 /Designer : Yu-tsen King

IAI 设计优胜奖

金禹岑 Yu-tsen King
中国台湾 Taiwan, China

2010 迄今 K+spacing 金玉岑空间设计公司 设计总监；
2005—2010 年 大匀空间设计公司 合伙人；
2000—2003 年 原相联合建筑设计师事务所设计师；
1998—2000 年 沃荷设计公司 设计助理。

Since 2010   Design director of K+SPACING ;
2005—2010   Partner of Symmetry Design ;
2000—2003   Designer of Aura Design Company ;
1998—2000   Design Assistant of Wohe Design Company.

台湾大哥大数字生活是台湾的电信业者透过概念店的形式，传达并呈现有别于一般 3C 产品的生活体验馆。通过科技互动的方式，提供创新的云端数字内容服务。基地位于台北市松山文创园区内，此区前身为台湾烟酒公卖局松山烟厂。1998 年停产后，2001 年由台北市指定为市定古迹。目前规划为台北文化体育园区，兴建体育、购物中心与文化展演空间等设施，成为台北市东区最具文化特点及创造活力的园区。早期的台湾，大榕树下常为人们茶余饭后交谈的聚集地，大家互相交换信息，分享生活。回顾历史，设计师想唤回过去人们坐在大榕树下彼此交流的温馨场景，以及运用设计将过去与未来的 3C 科技结合，营造最有人情味的体验场所。

Taiwan Mobile Digital Life is a concept store where people can experience different 3C products. By the way of interactive technology, it provides innovative Cloud digital content service. It's located at Songshan Cultural and Creative Park, Taipei City, Taiwan and this area was formerly known as "Taiwanese Provincial Tobacco and Alcohol Monopoly Bureau Songshan Plant". After production ceased in 1998, the plant was appointed by Taipei City Government as a cultural heritage in 2001. This area is planned currently for The Taipei Dome Complex and will be under the construction of sports space, shopping centers and cultural performance space to be the most cultural and creative area in eastern Taipei.In the early times in Taiwan, people often gathered under a big banyan to exchange information with each other and share their lives. The designer wanted to recall the warm sense of the site and combined the past with future 3C technology to create a place full of human warmth.

筑韵
LEICHT

项目地点：台湾省台中市西屯区河南路四段171号 / Location : No.171, Henan Road, Situn District, Taichung, Taiwan
项目面积：298平方米 / Area : 298 m²
公司名称：珥本设计 / Organization Name : Urbane Design
设 计 师：陈建佑 / Designer : Steven Chen

IAI 设计优胜奖

陈建佑 Steven Chen
中国台湾 Taiwan, China

URBANE DESIGN
陈建佑，毕业于中原大学室内设计系。
URBANE是温文尔雅的形容词，这样的形容也十分符合我们作品表达出来的氛围。
我们以"珥本"这两个字来表现思路脉络，珥为古代朱玉耳饰，玉乃自然优美之材料、耳则为人体构造极富细节之处，就像在设计的操作中，我们想呈现出材质最原始的特性，探讨其在空间中比例、分割、轻重、厚薄的关系，并经由空间分割排列的过程，达到功能与动线，光线与阴影，穿透与封闭的相对关系。

URBANE DESIGN
Graduate -Department of Interior Design, Chung Yuan Christian University.
The "suave and urbane" atmosphere is what we exactly want to express. URBANE has two meanings. "Er" is a kind of ancient jade earrings, and the jade is graceful product of nature, while ear is one of subtle human body, as we would like to display the original characteristics of material in the design, discussing the relationship of proportion, division, weight and thickness, the relative relation between function and dynamics, light and shadow, penetration and closure in the process of space division and arrangement.

本项目为挑高6米的钢结构建筑，空间的尺度与跨度成了表现对比性与自由配置的绝佳基础。结构上的缩放是以层递性的顺序来实现的，并将灯光、空调、音响、机电系统整合贴附于建筑物的内在皮层，就像厨具嵌入家具一般，使这些系统能自然地与空间结合，形成静谧的建筑语汇。其次置入厨具的尺度，让系统性的数据规划转为独特设计与配置自由的的德国劳斯美学，并强调空间布置的逻辑性与便利性，使德国制造的严谨性能合理且精准地被执行于每个购买者家中。

The base site is a steel structure building with ceiling 6 meters; spatial scale and span become an excellent basis performing comparison and free distribution. Using order of layer to deliver scaling on structure, to integrate lighting, air conditioning, audio, electrical and mechanical system for attaching them to the inner cortex of the building, just like concept to embed the kitchen furniture, which makes these systems naturally combined with space and form quiet building vocabulary. Secondly, into kitchen scale, to make systematic plan statistics being converted to LEICHT Aesthetics with unique design and free distribution, also strengthen logic and convenience of arranging space, which allow the rigor of made in Germany can be reasonable and precisely implemented in house of each purchaser.

中泰家具博览中心
Jongtay Furniture Expo Center

项目地点：广东省中山市西区沙朗高科技开发区沙港路1号 / Location：No.1, Harbour Road, High-tech Development Zone of Sand, Zhongshan City West Sirloin
项目面积：10 000 平方米 / Area：10 000 m²
公司名称：珠海天王空间设计有限公司 / Organization Name：Zhuhai Tian Wang Design Co., Ltd.
设 计 师：张成荣 / Designer：Zhang Chengrong

IAI 设计优胜奖

张成荣　Zhang Chengrong
中国　China

珠海天王空间设计有限公司总经理、首席设计师；
毕业于重庆师范大学首届摄影专业；
概念空间、建筑及景观首席设计师；
办公家具行业品牌整合策划资深推广人士；
曾任北京理工大学空间设计学院客座教授。

General Manager and Chief Designer in Zhuhai Tian Wang Space Design Co., Ltd.；
Graduated from Chongqing Normal University, the first professional photography speciality；
The chief designer in architecture and landscape concept space；
Brand integration marketing senior people in furniture industry；
He was a visiting professor at the Beijing university of technology institute of space design.

本案属于一个旧楼改造项目，由于原建筑结构横梁和柱体的错综复杂无法更改或重建，设计师结合设计主题大胆地利用建筑主体的不利因素变成设计的元素，用现代的手法将中泰的品牌文化赋予新的生命。中泰以"和文化"作为企业文化，"和生万象"是中泰品牌文化的精髓。在众多中国文化元素中，筛选出六合（六和）、万字回纹、荷花、格栅等最契合概念设计元素，大胆运用中国红作为主色，创造出震撼且富有感染力的视觉效果。设计上，重点在于运用现代形式及现代材料材质来表达中国传统"和"文化的内涵；通过VI企业形象与空间进行有机转换，做到品牌文化和空间设计的完美表达。

This design involves in reconstruction of an old building. The beams and extremely sophisticated pillars of the former building allow no changes or reconstructions, by taking reference from the design theme, the designer makes use of disadvantages of the building as design elements and brings new life to the brand culture of JONGTAY with modern techniques. JONGTAY takes "harmony culture" as the corporate culture, with "harmony brings everything" as the essence of JONGTAY's brand culture. Among diversified Chinese cultural elements, elements fitting best to conceptual design, including the six directions (six harmonies), svastika, lotus flower and grids are selected, together with the dominant hue acted by Chinese red, shocking visual effects full of infectivity are created. As for design, modern styles and modern materials are used mainly to deliver the connotation of the traditional "harmony" culture of China. Perfect expression of brand culture and space design is realized through the organic transition between VI corporate image and space.

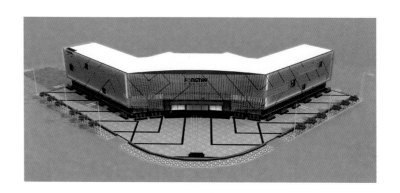

Hotel and Restaurant Space
酒店、餐饮空间

Hui Hotel Shenzhen	36
Macalister Mansion	38
Food House Restaurant, Huludao	42
Macau Praia Restaurant	46
Taste of Fashion —— Xi Xiangyu	48
NUR	50
Three Monkeys	52
Masu	54
Summer Invitation · Yuexian Fang Restaurant	56
YIYI Art Coffee	58
Taiyuan DaoFuLin Restaurant	60
Bao Bei Dan Restaurant	62
La Stazione	64

36	深圳回酒店
38	麦卡里斯特酒店
42	葫芦岛食屋
46	澳门海擎天茶餐厅
48	品味时尚——禧湘遇
50	NUR
52	三只猴子
54	Masu
56	夏宴·粤鲜舫
58	壹壹艺术咖啡
60	太原稻福临餐厅
62	宝贝蛋餐厅
64	老火车站

深圳回酒店
Hui Hotel Shenzhen

项目地点：深圳市福田区红荔路 3015 号 /Location : No.3015, Red Li Road, Futian District, Shenzhen
项目面积：8 800 平方米 /Area : 8 800 m²
公司名称：杨邦胜酒店设计集团 /Organization Name : Yang BangSheng & Associates Group

IAI 年度最佳设计机构

YANG 杨邦胜酒店设计集团
Yang BangSheng & Associates Group

杨邦胜酒店设计集团（以下简称YANG），是由著名酒店设计师杨邦胜及其董事共同创立，专业从事国际顶级品牌酒店与高端室内设计服务的大型设计公司，汇聚美国、意大利、法国、新加坡等数百名设计英才。YANG 严苛项目细节，追求完美，坚持建筑、室内、景观"一体化"的设计理念和机电、灯光一条龙的服务体系，获得（STARWOOD）喜达屋、凯悦等国际酒店管理公司的高度认可。
在 2014 年凭借《三亚海棠湾 9 号度假酒店》获全球最高酒店大奖美国 IIDA "全球卓越设计大奖之酒店设计奖"，并将在巴黎和纽约设立办事机构。

Established by the famous hotel designer Yang Bangsheng and his partners, Yangbangsheng & Associates Group (YANG) is a large-scale design company specializing in international top brand hotel design and high-end interior design. With over a hundred design talents from the USA, Italy, France and Singapore etc.. With unique creativity and outstanding services, the company is well recognized by various international hotel brands such as STARWOOD, ANANTARA etc.
YANG's versatile creativity has enabled it to win various design awards, including the much coveted Best of Hotel Category Award in the IIDA 2014 Global Excellence Awards Competition. It will eventually establish branch offices in Paris and New York.

回酒店是深圳首家新东方设计精品酒店，由多多集团与 YANG 杨邦胜酒店设计集团共同斥资并倾力打造。坐落于深圳市千米绿化中心公园旁，紧邻华强北，由旧厂房改造而成，并于 2014 年 7 月 11 日正式开业。酒店整体设计以新东方文化元素为主，并通过中西组合的家具、陈设及中国当代艺术品的巧妙装饰，呈现出静谧、自然的中国东方美学气质。大堂设计中，整面绿色墙植与一字排开的鸟笼，让空间化繁为简，加上清脆悠扬的鸟鸣，让人有"蝉噪林逾静，鸟鸣山更幽"之感。酒店整体用色和谐统一，结合用心调试的每一束光源，让空间散发出宁静、优雅的文化气质。

Hui Hotel is the first boutique art hotel in Shenzhen. Located in the most prosperous business circle, Hui Hotel used to be an old factory. Shenzhen Central Park nearby provides free natural landscape for this Hui Hotel. Although the total investment of the hotel only takes up less than 1/3 of the five-star hotel, it is the most distinctive hotel in this area. The design of the hotel focuses on new oriental elements and creates peaceful and natural Chinese aesthetic effect through a mix of Chinese and western furniture, as well as the smart decoration of displays and China's contemporary artworks. In the lobby, the whole piece of living wall and bird cages in a row make the space simpler. In addition, Hui Hotel seems to be more quiet while the birds are singing. The hotel applies harmonious colors and carefully adjusted lighting, which makes the space full of peaceful and elegant cultural ambiance.

麦卡里斯特酒店
Macalister Mansion

项目地点：马来西亚槟城 /Location : Penang, Malaysia
项目面积：1 700 平方米 /Area : 1 700 m²
公司名称：新加坡 MOD 设计公司 /Organization Name : Ministry of Design Pte., Ltd.
设 计 师：佘崇霖 /Designer : Colin Seah

IAI 最佳设计大奖

MINISTRY OF DESIGN.

新加坡 MOD 设计公司
Ministry of Design

新加坡 MOD 设计公司是由佘崇霖先生创立。公司的理念在于质疑惯例，对围绕在人们身边的空间进行重新的定义，试图用设计为生活赋予更多意义。在 MOD，他们喜欢逆流而上，而不只是设计的解决办法，并为他们的项目提供全盘的构想和体验。由此而产生的设计思路可以自然转化为各种设计方案和媒体传播手段：建筑、产品设计、室内设计、品牌策划、标识、景观，甚至可以将不同专业的内容汇集至一个项目中。

Ministry of Design was created by Colin Seah. MOD's explorations are created amidst a democratic studio-like' atmosphere and progress seamlessly between form, site, object and space. At MOD, They provide their clients services that transcend mere design skill sets or technical prowess. They prefer to start far upstream and instead of merely designing solutions, they design holistic experiences. The resultant design thinking then translates into a wide variety of possible downstream design applications and media: be it architecture, product design, interior architecture, branding, graphics, landscape or even the weaving of diverse disciplines into a single project.

麦卡里斯特酒店被一种奇怪的和世故的生活方式概念化，在这里，客人们都能经历到像在实际家中一样的殷勤款待。在这个封存了一百年历史的大厦中，麦卡里斯特酒店有六个部分：五个餐饮服务点和一个旅馆。每个部分都是整体的一部分，但每个空间都有自己的显著特点餐厅、小房间、蒲甘酒吧、客厅、草坪和八个房间。酒店名字是根据槟城的陆军上校诺曼·麦卡里斯特起的。为了与保守的大厦这样的背景形象对抗，设计师小心翼翼地去适应这继承下来的空间并保留了它的重要特点而且采用了同时期的设计，很好地平衡了对过去的怀旧和与未来的关联。以过去的同时期的奇怪风格来绘画，整合的不仅是存在的历史建筑的线索，还有诺曼·麦卡里斯特这个人物故事带有历史性的叙述……从某种程度上来说，历史本身被"适应性地重复利用了"。
麦卡里斯特酒店是马来西亚颇具设计感的酒店。它也因在《Conde Nast Traveler》上了 2013 年世界新酒店名单而名望大增。编辑花了 10 个月从 1000 个酒店中选出 154 个，麦卡利斯特酒店是 26 个收获到 3-flame 评级酒店中的一个。

Macalister Mansion is conceptualized as a quirky and sophisticated lifestyle destination where guests are treated to experiences similar to those found in the hospitality of an actual home.Housed in a 100-year-old historic mansion that has been conserved and adapted, Macalister Mansion consists of 6 entities: 5 Food and beverage service point and a hotel. Each entity can be enjoyed as part of the overall experience but are branded as distinct spaces that are typically found in a mansion - Dining room, Den, Bagan Bar, Living room, Lawn and Eight rooms. The name Macalister Mansion honours Penang's British Governor Colonel Norman Macalister. Set against the backdrop of a conserved mansion, the heritage spaces have been carefully adapted and key features conserved and infused with contemporary design, striking a good balance between the nostalgia of the past and a vision of relevance for the future.Drawing from the past in a quirky, contemporary way, MOD integrated not only the existing historic architectural cues, but also the tale of Norman Macalister as an underlying quasi-historical narrative... in a way, history itself has been "adaptively reused". Macalister Mansion has recently made the design cut to join Design Hotels and is currently the only one in Malaysia. It has also made Conde Nast Traveler's prestigious Hot List of World's Best New Hotels, 2013. The editors spent 10 months selecting the top 154 from 1000 hotels, Macalister Mansion was 1 of 26 hotels to receive the full 3-flame rating.

葫芦岛食屋
Food House Restaurant, Huludao

项目地点：辽宁省葫芦岛市滨海路7号 / Location : No.7, Binhai Road, Huludao, Liaoning
项目面积：2 101 平方米 / Area : 2 101 m²
公司名称：大连纬图建筑设计装饰工程有限公司 / Organization Name : V2 Gether Architectural Decoration Engineering Co., Ltd.
设 计 师：赵睿 / Designer : Zhao Rui

IAI 最佳设计大奖

赵睿 Zhao Rui
中国 China

现任大连纬图建筑设计装饰工程有限公司CEO、设计总监；
2008年收购大连中信建筑装饰工程有限公司，后更名为大连纬图建筑设计装饰工程有限公司至今；
负责设计项目涉及国家大型建筑规划工程建设、酒店工程、公共住宅、私人会所、别墅、厂房、展厅、办公设施等；
1997年成立北京正中建筑装饰工程有限公司。

CEO, Design director of V2 Gether Architectural Decoration Engineering Co., Ltd.;
2008 Acquire Zhongxin Architecture Decoration Engineering Co.,Ltd.;
Rename V2 Gether Architectural Decoration Engineering Co., Ltd.;
Programs involve National large construction planning project construction, Hotel, Public house, Private club, Villa, Factory building, Exhibition Hall, Office Facilities etc ;
1997 Establish Beijing Zhongzheng Architectural Decoration Engineering Co., Ltd.

食屋项目前身是典型七八十年代复古建筑风格，地理位置相对优越，视野开阔，窗外直面无边海景。它定位为私人会所，供主人和亲友在此聚会之用。稻草是单体瘦弱的，但当它形成一片，并不断复制、阵列分布时，在整体上便会呈现出一种非均衡的力量感。设计师尝试把"稻草"具备的基本精神置换成一种空间构筑语言融入到食屋整个空间的设计中去，进而出现了入口前厅的"稻草"装置，以及在每个空间节点的墙身和顶棚上延续的不规则木条肌理。当空间被"稻草"所形成的灰色调的背景完成后，其他灵动的空间节点便是靠一些质感丰富的物件摆设来担当角色，进一步烘托空间气氛。

"Food House" project is a typical vintage architectural style of the 70's and 80's. The project enjoys a relatively favorable geographic location and an extensive view with the windows facing the boundless ocean. "Food House" is a private club for gatherings of owners, relatives and friends.Straws are extremely common and simple plants, which seems kind of weak instead of strong, but when they gather together and constantly duplicate the distribution of arrays, as a whole they will show an unbalanced strength. The designers try to displace the basic spirit of straws into the narration of whole space, and thus the straws appear in the entrance lobby, on the wall body of every space node and on the ceilings' irregular wood texture. When the space is coated with the gray background formed by straws, other dynamic spaces nodes play a role in presenting atmosphere by some decorations with fine texture.

Hotel and Restaurant Space 酒店、餐饮空间

设计师推崇对日常物品的价值进行探索和挖掘，最终让它们产生一种新的结构关系。并试图在"食屋"空间中的体现极具差异性的物件与空间的共融。与木色反差较大的玻璃工艺灯，路边捡来的枯枝和食用后的贝壳等物件经现场再创作形成的立体浮雕墙，包括桌上的白色碟子和透明高脚杯，市场上淘来的小葫芦等空间里的物件呈现出一种透气的整体感。

The designer advocates a kind of spirit, which is the process of exploring the values in common and easy objects to produce a new structural relationship in the end. And attempt to get to this state by the communion of the objects of great diversity and space in Food House: objects like glass craftwork chandeliers in sharp contrast with wool in color, dead branches collected in the street, and shells after eating can be recreated into a three-dimensional relief wall. Besides the wall, all objects in the space, including the white plates, transparent goblets on the table and small gourds picked up on the market, etc. present a breathable sense of integration.

澳门海擎天茶餐厅
Macau Praia Restaurant

项目地点：澳门林茂大马路 /Location : Linmao Road, Macao
项目面积：200 平方米 /Area : 200 m²
公司名称：宁波博洛尼装饰公司 /Organization Name : Boloni Decoration Company
设 计 师：范业建 /Designer : Dave Fan

IAI 设计优胜奖

范业建　Dave Fan
中国　China

2005—2008 年　北京主题建筑事务所，室内设计师；
2008—2011 年　珠海上端装饰公司，室内设计总监，创始人；
2011—2014 年　宁波博洛尼装饰公司，室内设计总监。

2005—2008　Beijing Zhuti Architects/Interior Designer ;
2008—2011　Zhuhai Shangduan Company/Founder and Interior Design Director ;
2011—2014　Boloni Decoration Company/Interior Design Director .

此项目坐落在澳门林茂大马路海擎天，邻近澳门游艇会、澳门购物中心、十六浦及拱北口岸，繁华之处，可见一斑。业主希望能够为年轻的消费者提供一个自然而舒适的休闲就餐空间．空间的概念主要用抽象的设计元素来表达自然主义的主题，聆听大自然的呼吸，感受大自然的静谧。抽象树状的墙面造型，黑色镜面的背景，彰显着原始森林般的深邃；火烧面的墙面石材，有着山石一样的质感，硬朗略显粗糙。一层入口右侧的水银镜丰富了空间的进深感，似瀑布般从屋顶泻至地面，在二层空间的顶面飘过几朵洁白的浮云，仿佛有一丝微风吹过。在走廊的尽头是烤漆玻璃制作的绿树剪影，拉伸了空间的视线和进深。草绿色的餐桌像是穿梭于黑色森林的小精灵，使整个空间承载着生命和希望。

This project is located in Avenida Marginal do Lam Mao—the Praia Community and this district is very prosperous, there being Macou yacht club, Macau shopping center, Ponte 16 and Gongbei Port in the surrouding area. The owner hopes to provide a natural and comfortable dinning space for young consumers. The design idea of the space is expressed by abstract design elements to voice the theme of naturalism. Once you are here, you can listen to Nature and feeling Nature. The metope modeling of abstract trees and the background of black mirrors give you a sense of deeper thinking from virgin forest. Also, the flamed stone materials possess the sense of a rock, being strong and somewhat rough. The mercury mirror on the right of the first floor entrance enriches the spatial depth, like the waterfall cascading off the roof. Some spotless and white clouds drift over the roof of second floor, it seems that a breath of wind is brought. The trees silhouette made by paint glass at the end of the corridor stretch the visual images and spatial depth. In addition, Grass green dinning-tables are decorated in the space that looks like a lovely fairy shuttling in Black Forest, conveying the life and the hope.

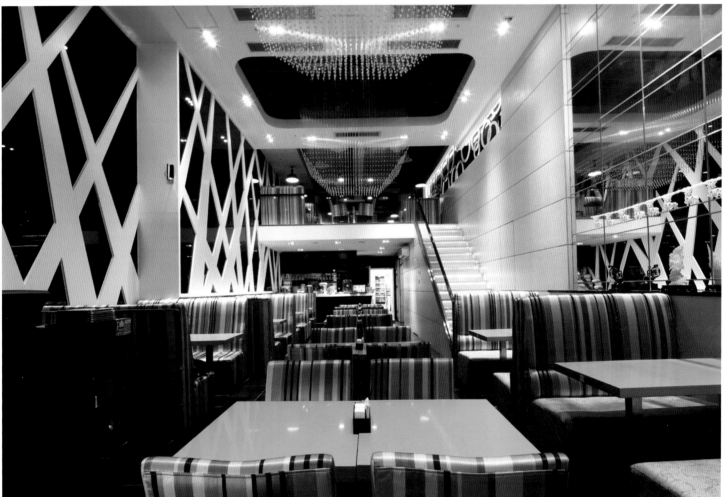

Hotel and Restaurant Space 酒店、餐饮空间

品味时尚——禧湘遇
Taste of Fashion —— Xi Xiangyu

项目地点：湖南省长沙市新世界百货六 /Location : 6F, New World Department Store, Changsha, Hunan
项目面积：550 平方米 /Area : 550 m²
公司名称：上海亿端室内设计有限公司 /Organization Name : Shanghai Yiduan Interior Design Co.,Ltd.
设 计 师：徐旭俊 /Designer : Xu Xujun

IAI 设计优胜奖

徐旭俊　Xu Xujun
中国　China

上海亿端室内设计有限公司首席执行创意总监；
国际注册高级室内设计师；
中国室内设计师协会专业会员；
2012—2013 年中国室内设计师年度封面人物。

Chief Executive Officer Shanghai Yiduan Interior Design Co.,Ltd ;
Senior Interior Designer International Accreditation and Registration Institute ;
Professional Member of China Interior Designers Association ;
2012—2013 The cover person of China Interior Designers Association .

禧湘遇餐厅的设计灵感来源于湘文化的吊脚楼，采用下沉和抬高的表现手法，使整个空间层次错落有序，给人留下湖南本土吊脚楼的建筑印象。设计本着绿色环保性、原创趣味性、文化艺术性、人性化设计等原则进行设计。在材料运用上主要采用素水泥、旧木板等环保材料。在设计上强调高低错落的震撼力和冲击力的同时，又注重研究平面、立体、色彩三大构成在整体空间上的协调统一。家具的配置注重实用、个性和美观。灯光格调偏向高雅、自然。布局上打破"车排式"的传统方式，用围合式的吊脚楼，结合时尚的装置艺术和混搭的装饰风格，让来访客在这个空间中就能感受到时尚的湘菜文化。从主入口到整个空间过道动态线的合理布局，再到水、花和石等景观的点缀，皆体现着人与自然和谐相处的浪漫情怀。

Jubilee in Hunan restaurant design inspiration from the Hunan culture Diaojiao Lou,clever use of space layout and dining atmosphere, the sinking and elevated expression, make whole space level of scattered and orderly, give a person leave Hunan native Diaojiao building impression.The design is focused on Green venvironmental protection,original interest,culture and art.Representative materials:cement, old boards etc.The design space strongly scattered high and low, give a person with strong shock force and visual impact. Furniture, pay attention to the practical character of original type, beauty and interest based. Lighting design in favor of elegant, natural style.To break the "traditional layout cars", use Weiheshi Diaojiao Lou style, combined with fashion art installation and mix of the decorative style,let the customers can in the space to feel this is the fashion of Hunan culture restaurant.The reasonable layout of the space dynamic line from the main entrance hall, and then to the waterscape, landscape, flowers, dried bamboo scenery and stone scenery landscape ornament, as if in the interpretation of a harmony between man and nature of the touching story, and romance.

Hotel and Restaurant Space 酒店、餐饮空间

NUR
Nur

项目地点：香港 /Location : Hong Kong
项目面积：204.36 平方米 /Area : 204.36 m²
公司名称：宁设计事务所 /Organization Name : J. Candice Interior Architects
设 计 师：陈浩宁 /Designer : Candice Chan Chao

IAI 设计优胜奖

陈浩宁 Candice Chan Chao
中国香港 Hong Kong, China

陈浩宁，香港出生。
在美国雪城大学深造室内设计，以优异成绩取得
Bachelor of Fine Arts 学士学位。
在纽约工作期间，参与百老汇剧院修复、纽约著
名 92 地标工程，并参与多间当地跨国公司写字
楼设计工作；
2009 年回港，自立门户，创立宁设计事务所，
提供室内设计及建筑专业服务，著名客户如大家
乐、意粉屋、精品餐厅及酒吧等。

Candice Chan Chao, born in Hong Kong, China Studied in the Syracuse University, and received Bachelor of Fine Arts Degree with honors；
Graduating summa cum laude with a B.F.A from Syracuse University, she spent her post-college years in New York, resurrecting a Broadway theatre, overhauling studios and offices for multi-national corporations and developing interiors for New York's famed 92Y；
After returning to Hong Kong in 2009, Candice headed her own professional firm, J. Candice Interior Architects, designing rescommercial interiors for notable brands like Cafe de Coral, The Spaghetti House, bars, clubs and office around the globe.

NUR 主厨 Nurdin 深信，最好的食物的食材一定要新鲜。为确保食材新鲜，避免食物在运输过程中洐生碳足迹，设计师特别打造了一个面积为 100 平方米的空中花园，种植自家原材料，自给自足。NUR 的创作灵感源自实验室，透过不同形状的玻璃量杯和大小不一瓶子，演绎烹饪美食学。空间设计以浅灰色和白色为主调，并以钢铁作点缀，浅灰色的木地板配合自然的白色墙身，营造出独特的气氛情景。

A 100 meeditable garden with an aim to sufficiently self-supplied on self-grown ingredients and herbs, the idea is to create a "sky garden" in the city. While Nurdin collaborates with local farmers and food suppliers to ensure his ingredients are at their freshest condition when arrived to the restaurant. Beside green ideas, the design for NUR is noticeable for its sophisticated contemporary design inspired by laboratory theme. From the jar of fermented ingredient to glass beakers and jars, each element is custom design to reflect the spirit of gastronomy. The tone inside the restaurant is generally light grey and white with occasion use of steel and iron. Combinations of natural grey wood floor with white wall compliment with the natural greens and herbs, create a truly unique and inspiring venues.

三只猴子
Three Monkeys

项目地点：香港 /Location : Hong Kong
项目面积：204.39 平方米 /Area : 204.39 m²
公司名称：宁设计事务所 /Organization Name : J. Candice Interior Architects
设 计 师：陈浩宁 /Designer : Candice Chan Chao

IAI 设计优胜奖

餐厅名源于一个关于三只猴子的古老神话故事——三只来自外太空的猴子,搭乘自己创制的飞船,来到地球,初尝啤酒,从此决定留下,并在香港建立了一个可以与人分享啤酒滋味的地方。因此,餐厅设计模拟宇宙飞船,店内采用悬吊的灯饰,黑色铸铁制的餐椅,没有特别打磨的餐桌面,挂着日本佳酿黑白片的墙身,令顾客在品尝美酒之余,更能享受铁板串烧的原始风味及感觉。

Based on a story of three monkeys inhabited Mars. These three monkeys travelled to planet Earth for expedition and discovered something spectacular "Drafted beer". Ever since, they left their spaceship and . Their mission is to craft the finest drafted beer and offer to everyone in the heart of Hong Kong. Concrete flooring, raw table top and cast iron structure are other major elements to craft the rustic industrial vibe from the comic. At bar area, bottles of Japanese sake are showcase in a copper plated cabinet to highlight from other drinks. Additionally, pictures of black and white Japanese vintage photos and graphics are mounted on brick wall adds a quirky personality to the venue.

Hotel and Restaurant Space 酒店、餐饮空间

Masu
Masu

项目地点：香港 /Location : Hong Kong
项目面积：167.23 平方米 /Area : 167.23 m²
公司名称：宁设计事务所 /Organization Name : J. Candice Interior Architects
设 计 师：陈浩宁 /Designer : Candice Chan Chao

IAI 设计优胜奖

封建时期的日本，Masu（枡）是一个度量人民每天米量的方形木盒，餐厅业主希望顾客在享受美食时，能够体会到当年日本人对每日食物的要求及执着。与此同时，餐厅主厨坚持采用新鲜食材及传统烹饪手法，只为顾客带来真正地道的日式美食，如寿司及炉端烧。

Since feudal times in Japan, Masu was a wooden box used to measure rice, enough to feed one person for the day. The owner of the restaurant wish to evolve this concept to provide their customer with true taste of Japan for one meal. Shedding all unnecessary flourished, Masu's chef Kazuya Nemoto emphasised on using only the freshest local and regional ingredients, with traditional Japanese cooking techniques including Kobatayaki, sushi and other specialties.

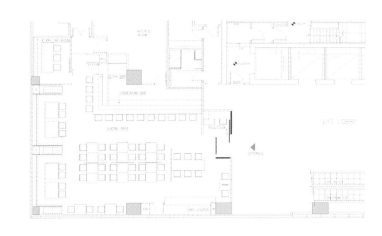

Hotel and Restaurant Space 酒店、餐饮空间

夏宴·粤鲜舫
Summer Invitation · Yuexian Fang Restaurant

项目地点：浙江省杭州市余杭临平南苑街 103 号麦道大厦三楼 /Location : 3F Building Maidao, 103 Nanyuan Road, Linping District, Yuhang City, Hangzhou, Zhejiang
项目面积：2 500 平方米 /Area : 2 500 m²
公司名称：中国美术学院 /Organization Name : China Academy of Art
设 计 师：王海波 /Designer : Wang Haibo

IAI 设计优胜奖

王海波 Wang Haibo
中国 China

2010 至今　浙江亚厦股份副总设计师；
中国美术学院国艺城市设计研究院副院长；
第九研究院院长；
2003 至今　中国美术学院讲师；
2004—2009 年　浙江中和建筑设计有限公司设计总监；中国建筑装饰协会高级室内建筑师；高级景观设计师；浙江省创意设计协会理事长；
1999—2003 年　中国美术学院环境艺术专业本科。

Since 2010,Deputy Chief Designer in the mansion co., LTD., zhejiang province,vice president of the urban design and research institute,the China academy of fine arts skill;the chair professor of the ninth academy ;
2003 until now,Lecturer in China Academy of Art;
2004—2009　Design director of Zhejiang Zhonghe Architecture Design Ltd;Senior interior architect of China Building Decoration Association ;Senior Landscape Designe;Council President of Zhejiang Creative Design Association ;
1999—2003　Bachelor degree, major in environment art design in China Academy of Art.

夏夜，萤火虫漫天轻舞、流水叮咚、竹影摇曳，营造的不只为享用美食时的那一刻心境，更带回久远的乡村记忆。餐厅兼顾早茶、中餐及婚宴功能。用现代的手法、质朴的材料营造浪漫、温馨的主题餐厅。主要材料为仿旧花岗石、毛竹、竹板、青砖、角钢、不锈钢、玻璃。

The Design Description of Yuexian Fang Resturant,Hangzhou: At a summer night,the fireflys are flying everywhere. The stream is flowing and the bamboos are waving to the wind, which not only creates the mood to enjoy the cousin,but also brings the memories in the countryside long long ago.They restaurant caters for Cantonese morning tea, Chinese meal and wedding feast . It's a theme restaurant which boasts of a romantic and cosy atmosphere because of its simple material and modern structure. Main material:faded granite, moso bamboo,bamboo clappers, black brick,angle cleat, stainless steel, glass.

壹壹艺术咖啡
YIYI Art coffee

项目地点：福建省厦门市湖里区湖里大道 10-12 号文创园 4 楼 2 层 /Location : 2F, Building 4, No.10-12 Huli Road, Huli District, Xiamen, Fujian
项目面积：158 平方米 /Area : 158 m²
公司名称：厦门方式装饰设计有限公司 /Organization Name : Fancy Design Co., Ltd.
设 计 师：方国溪 /Designer : Fang Guoxi

IAI 设计优胜奖

方国溪　Fang Guoxi
中国　China

2008 至今　厦门辉煌副总经理、设计中心总经理；
2006—2008 年　厦门辉煌设计部经理；
2001—2003 年　厦门辉煌设计部设计师创办方式设计机构；
1998—2001 年　自由设计师。

Since 2008,Vice President of Xiamen Huihuang Design Department,General Manager of Design Center;
2006—2008　Manager of Xiamen Hulhuang Design Department;
2001—2003　Designer of Xiamen Hulhuang Design Department;
1998—2001　Freelance designer.

壹壹艺术咖啡是厦门第一家艺术与咖啡相结合的场所，融合艺术作品和生活情调为一体，提倡原创、健康的生活理念。约 160 平方米的空间有着整面墙壁的竹片，午后的阳光透过竹片落在了最原始的水泥地板上，木质的桌椅，棉布的沙发，生机盎然的绿色植物，极富美学设计的小饰物、充满古典气息的茶壶和琉璃……映衬着柔和的灯光和舒缓的音乐，显得温暖而雅致。正如它的创办人方国溪所说的，"我们希望它朴素、大方、自由，不带任何的矫情和修饰"。竹片的选择源自于壹壹对健康、新鲜的食材和"健康、自然、本质"的追求。在这个长条形的空间里，创意性的靠窗栏面制作了半环绕的整面竹片，或长或短，不求定势的自由弯曲，显示了它的自然和韧性。原始的毛石、腐锈的钢板，也展示了空间朴素的质感、本真的情怀。

"YIYI Art coffee" is the first place that combined with art , brighting the literary flavor in Xiamen,and combines the art with life, promotes original healthy lifestyle.The coffee bar is more than 700 square meters, the wall is surrounded by the bamboo。The afternoon sunshine passes through the bamboo and falls on the most primitive cement floor, the wooden tables ,chairs, cloth sofas, the life plants, the aesthetic designed small accessories,the glass which is full of classical atmosphere ,the sunshine silhouettes against the soft light music,that seems so warm and elegant. As its founder Fang Guoxi said "we hope he is simple and generous, without any delicacy and modification. "
Why they choose the bamboo because the starting is health , nature, essence. In this strip shape space, creative surface made the all bamboo half round by the wall, or long or short, to set free bending untime , showing its nature and toughness. The original rubble, rusty steel which also shows the simple texture, true.

太原稻福临餐厅
Taiyuan DaoFuLin Restaurant

项目地点：山西省太原市南内环街189号 / Location : No. 189,South Inner Street, Taiyuan, Shanxi
项目面积：930平方米 / Area : 930 m²
公司名称：杭州境致装饰设计有限公司 / Organization Name : Hangzhou Jingzhi Space Design Co., Ltd.
设 计 师：魏志学 / Designers : Wei ZhiXue

IAI 设计优胜奖

魏志学　Wei Zhixue
中国　China

高级室内设计师，设计总监；
2008年　创办境致空间设计事务所；
2004—2008年　在杭州一所建筑设计事务所工作；
2003年　毕业于沈阳工业大学室内设计专业。

Senior interior designer,Design director;
In 2008 He founded in Jingzhi Space design firm;
2004—2008 He worked in architectural design firm in Hangzhou.;
In 2003 He graduated from shenyang university of technology professional interior design.

在太原已经小有名气的时尚餐饮品牌的新店中，设计师将红地毯、T台作为设计的重点。
顶面裸露的管道，灰色的墙面涂料，现浇的水泥地面，被大面积的斜拼铜板"包裹"的水吧及卡座，还有楼梯旁大面积的植物墙，在满足正常照度下，在光源上尽量的做"减法"。
从一楼进门开始直至餐区，贯通了一条"蜿蜒"而上的红色防滑玻璃发光T台。由T台自然分隔了两大两小，四个主餐区。整个餐区的平面、顶面、地面、立面上的大块面处理，都是为了在空间表现中，后退一个层次，来烘托整个发光T台。这个就是设计师在阐述的"平民化"设计。

Dsigners focus on the "Ored carpet", "T station "in this new stylish restaurant and minor celebrity in taiyuan.
Top of the bare pipe ,gray metope paint,cast-in-situ concrete floor,is a large area of oblique spell copper "package"of the water,clear file and gets stuck,and staircases plants in a large area of the wall.Under normal illumination and meet also try to do "subtraction"on the light source.
Startinng from the first floor door,until the dinning area,through a "snack"and on the red glow anti-skid glass catwalk.From the catwalk natural divide the two big and small,four main metals area.The large surface processing on the flat,surface,floor and facade of the whole food division is to highlight the whole luminous catwalk as the background in the space performance,which is the so-called "civillian design".

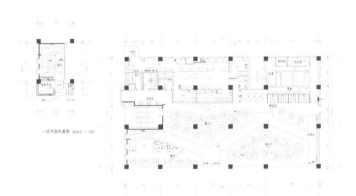
一层平面布置图 SCALE 1:100

宝贝蛋餐厅
Bao Bei Dan Restaurant

项目地点：浙江省杭州市运河上街 / Location : Yunheshangjie (UP), Hangzhou, Zhejiang
项目面积：860 平方米 / Area : 860 m²
公司名称：杭州境致装饰设计有限公司 / Organization Name : Hangzhou Jingzhi Space Design Co., Ltd.
设 计 师：魏志学 / Designers : Wei Zhixue

IAI 设计优胜奖

宝贝蛋餐厅是几位服装业从业者的跨界产物。"有所联系，能够让人一进餐厅就能感受到，这个是做服装的人做的餐厅。"
公共区域顶面悬挂数以万计的锈蚀圆管，局部以水波纹图案留白。与餐区内的顶面锈蚀钢板镂空的发光灯珠水波纹造型呼应。走道的图案硬包与餐区的锈蚀钢板立体造型呼应。8毫米厚钢板线切割点阵造型隔断，16毫米的锈蚀圆管不规则排列隔断，彩色线轴的排列隔断。地面的3个色号，深浅搭配的水泥砖立体图案地面……以点成线，以线成面，点、线、面组合成空间。不断强调构成关系，空间与服装的元素结合。时尚跨界，大抵应是如此。

"Bao Bei Dan" is the cross product of several uncles' & aunts' from clothing industry. "Contact has been able to let people feel that this is a restaurant operated by a person doing clothes business as soon as they enter into the restaurant." Tens of thousands of corroded pipes are on the top of the public area, with water ripple pattern blank. Dining area and the top surface rust steel hollow beads glowing lights ripple shape echoes. Walkway pattern hard pack and three-dimensional modeling of corrosion steel dining area echo. 8mm steel wire cut off dot shape, rusting pipe 16mm irregular arrangement partition, partition arrangement colored spools. Ground color number three, with a depth of three-dimensional pattern of cement brick floor …… To point to a line, the line into the surface, a combination of spatial points, lines, polygons. Repeatedly stressed that a relation, combining elements of space and clothing. Cross fashion, probably should be the case.

Hotel and Restaurant Space 酒店、餐饮空间

老火车站
La Stazione

项目地点：上海市静安区延平路 98 号 /Location : No.98, Yanping Road, Jinan District, Shanghai
项目面积：250 平方米 /Area : 250 m²
公司名称：DL 建筑 /Organization Name : DLArchitecture
设 计 师：多米蒂娜·莱普 /Designer : Domitilla Lepri

IAI 设计优胜奖

多米蒂娜·莱普　Domitilla Lepri
意大利　Italy

建筑师、室内设计师；
2006 年　在上海市中心成立了分公司，该分公司拥有一支国际型建筑师和室内设计师团队，多元文化在他们的项目风格上得到很好的体现；
2002 年　成立上海多米装潢设计有限公司，项目大多数是罗马和米兰旧公寓翻新工作；
2000—2002 年　多米蒂娜任职于罗马 Desarco 公司，专门从事罗马周围的公寓翻新和别墅建设工作；
2000 年　荣获罗马大学建筑学学位，专攻建筑修复和历史古迹保护研究。

Architect, Interior designer ;
In 2006 Domitilla Lepri moved to Shanghai, opened her branch office in the heart of the hustling bustling Shanghai, where she still lives and works. The office includes international team of architects and interior designers.;This unique team characteristic creates a blend of different cultures, which is reflected on the style of their projects;
In 2002 Domitilla formed her own company "DLArchitecture & Interior Design", managing projects in Rome and Milan. Most of the projects were renovation of old apartments located in downtown Rome and Milan ;
Between 2000 and 2002 she worked at "Desarco" firm in Rome, focusing in apartments renovation and construction of villas around Rome ;
In September 2000 Domitilla has been admitted to the Italian Professional Association of Architects (Ordine Degli Architecetti di Roma).

本项目灵感源自一座体现新艺术风格的老火车站。这座两层的场所呈现出典型 19 世纪末欧洲建筑风格。在建筑表面，优雅和衰弱是主导，主要元素有树枝形状的铁柱、隔断花墙、大幅马赛克拼图，墙上大量做旧的水泥砖、局部石膏镶嵌的顶棚，同时，许多细节元素如时钟、行李箱、装饰镶板，把顾客们带回了铁路交通的黄金时代。
一楼和二楼的连接由特制的新艺术玻璃杯架来完成，整个装置从一楼穿过打凿的空洞一直连到二楼顶棚。

This project takes her inspiration from an old art nouveau railway station.Located on two floors this venue has the typical flavor of a European architecture from the end of the nineteen century. On the facade elegance and decadent feeling leads,the main elements are iron pillars looking like tree branches, floral style partition walls, big mosaic composition, and a lot of broken old looking cement tiles on the walls and chalk tiles on selected area of ceiling. Many elements such as clocks, luggages and decorative panels bring the client back into the golden era of railways.
The continuity between the 1st and the 2nd floor is assured by the spectacular art nouveau glass holder which goes from the first floor through the void, up to the second floor ceiling .

所有这些创作的构想都与一座火车站息息相关，接待处像一个老售票亭，柜台底部是旧行李箱，墙上挂着的画作是曾经在火车上使用的终点指示牌，售票处马赛克方向拼图、等候区、月台、时钟、节日广告海报等都增强了人在途中和通向下一站迷人站点的感觉。

The creative ideas are all linked to the connection to a railway station. The reception desk looks like an old ticket booth, and the bottom of the counter is made with old luggages. On the walls, paintings representing old metallic plates with the indication of destination, once used on the trains. Mosaics decorations with directions on ticket office, waiting area, platforms, clocks, and advertisement poster on holiday's destination, all those elements contribute to the feeling of being on process of leaving to an amazing destination.

Entertainment and Clubs Space
娱乐、会所空间

First Thought & Creation	68
Bump Restaurant & Bar	72
Dekang Club	76
ABRAZO	78
LOST Lounge Bar	80
Yin Lang KTV	82
Shichen Fortune Club	84
State Banquet Reception Center	86
Master Film Studio in the Rim of Shanghai University	88
Long Shenghui Club	90

68	起初 & 创造
72	凹凸餐吧
76	得康会所
78	拥抱
80	迷失酒吧
82	音浪 KTV
84	世辰财富会所
86	国砚接待中心
88	环上大影视园区大师工作室
90	龙圣汇会所

起初 & 创造
First Thought & Creation

项目地点：台湾省云林县 /Location : Yunlin, Taiwan
项目面积：66 平方米 /Area : 66 m²
公司名称：鼎睿设计有限公司 /Organization Name : DINGRUI Design Studio
设 计 师：戴鼎睿 /Designer : Tai Ding-rui

IAI 最佳设计大奖

鼎睿设计有限公司
DINGRUI Design Studio

公司一路走来十几个年头，设计师用心、认真地看待每个案子，希望每个案子都可以表达出真、纯、净、沉、极的初心。简单、自然、自在、唯一、无所顾虑的心，作品注重人与空间之情感．

原名戴晴峰空间设计有限公司；
1998 年成立迄今，于 2008 年 1 月更名；
设计总监，戴鼎睿（Ray）。

The company over ten years, designers treat every case so seriously ;
They hope every case can show the willing of true,pure,calm,extreme.They want to show heart of the simple ,nature ,easy,only. The works pay more attention on the emotion between people and space;

Old name is Dai Dingfeng space design company ,change the name in 2008.1;
Design Director Ray.

一个校长的家园，一开始的想法是拥有一个可与三五好友聚会的地方，能一起谈天一起品茶读书的场所。在了解了当地的朴实感与热情后，设计师希望人们可以回到最初——那就是人的五感六觉，感受风动、土味、草香与木质的洗礼。荔枝树在旁陪伴着，看似简单却拥有所有。人脸上的微笑，被一切的感动与关怀包围，用原有的肌理与元素表达出真诚的待客之道。

Rinnai,Yunlin Home of a school principal. The initial idea is to have a place where gets together-chatting, reading or drinking with friends. After understanding the simple and enthusiasm of that area,designers hope people could back to the original sense to feel the wind, earth and smell the grass and wood. A hundred year's litchi tree is next to you and also accompanying with happiness. It seems so simple but actually you have everything. Smiles in people's face are surrounded and touched by every concern with the sincerely hospitality of original textures and element.

Entertainment and Clubs Space 娱乐、会所空间

Bump Restaurant & Bar

凹凸餐吧

项目地点：广东省广州市番禺区 /Location : Panyu County, Guangzhou, Guangdong
项目面积：900 平方米 /Area : 900 m²
公司名称：锐意（广州）设计 & 共和都市（香港）设计 /Organization Name : Ray Evolution Design Ltd&Republican Urban Design Ltd.
设 计 师：黄永才 /Designer : Ray Wong

IAI 最佳设计大奖

黄永才 Ray Wong
中国 China

锐意（广州）设计有限公司创办人、首席设计师
共和都市（香港）国际设计有限公司创办人、首席执行官。

Chief Designer and Founder of Ray Evolution Design Ltd.;
Chief Executive Officer and Founder Republican Urban Design Ltd.。

在城市化进程逐步加快的当下，凹凸餐吧试图以工业文明的肌理作为城市记忆的延续。

凹凸餐吧以后工业景观来贯穿整个设计环境主线——"立"于阵列松木的屏风，"破"于对角线的吧台。本案以锈蚀钢板的粗犷，废弃的碎木自然肌理来重拾工业文明的记忆碎片。阵列的竖向松木条倒影在环氧树脂水泥板地面上，把过于低沉的原有建筑顶棚虚化的同时，强调了竖向阵列的通透性与私密性。吧台顶的竖向条形肌理玻璃给顶棚"罩"上一层颇有玩味的模糊与暧昧，同时也虚化了耐候钢的笨重。

凹凸餐吧在选材上是以氧化过程其变化"有生命"的耐候钢、被废弃的碎木、自然肌理的麻石等再生环保材料为主。

Nowadays, as the accelerating Urbanisation rate taking places. Bump Restaurant & Bar (Bump) themed "industrial civilisation" as the designing texture that aiming for lasting city memories.

To memory the great industrial time, Bump adopt the concept of "post-industrial" as the principal line for environmental designing. "Stand" pine screen to be set an array; "Break" the diagonal of this array by placing the bar. Materials been chosen rusty steel and grinded wasted wood to represent tough and boundless for re-memorising what's gone away from us. Pine wood made for the screens has its reflection on epoxy resin furnished rough concrete floor, it emptied the feel caused by lower ceiling and meantime emphasised the see-through factor but with privacy remained which all comes from the vertical pine wood.By the bar, a pillar full of numbers of different Times Magazine cover pages under a huge fuzzy glass brings some interesting and a sense of flirtation.

For the materials used in Bump, they are all "alive" which representing life: already been oxidated corrosion resistant steel (oxygen), pine wood and waste wood, granite with its nature texture , some recycle & environmental friendly materials, etc.

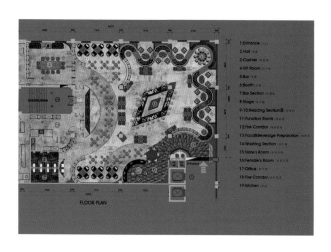

FLOOR PLAN

Entertainment and Clubs Space 娱乐、会所空间

得康会所
Dekang Club

项目地点：广东省广州市白云区 /Location : Baiyun District, Guangzhou, Guangdong
项目面积：2 000 平方米 /Area : 2 000m²
公司名称：锐意（广州）设计 & 共和都市（香港）设计 /Organization Name : Ray Evolution Design Ltd. & Republican Urban Design Ltd.
设 计 师：黄永才 /Designer : Ray Wong

IAI 设计优胜奖

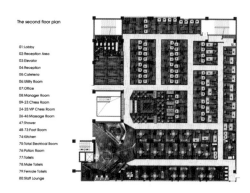

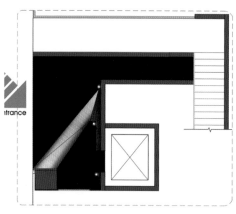

得康会所是坐落于广州尚佳广场二楼的一个休闲娱乐商业项目，经营面积2000平方米，主要消费群体是城市工薪阶层，供娱乐、聚会之用。得康会所的设计理念以"城市景观"为基本线索，以解构中国唐代的青绿山水画为基本出发点，在现代都市生活的诉求上，得康会所以独特的视角作出了回应。

因该项目在电梯间到接待大堂平面布局上，人流动线是本案的基本切入点，曲折蜿蜒的人流动线，宛如中国画的深山幽谷、白云萦绕，行人游赏、穿行其间。立面上的三角切割面各具姿态，增添行人乐趣与好奇。从接待大堂到自助餐厅，再到被服务空间动线上是由动到静的关系。接待大堂分别在两处入口和自助餐厅入口放置了三棵鸦青的枯树，古语云："山水以树始。"即说树是一幅山水画的开始，在这里对整个大堂接待空间起到抽象标示的作用。

Club is a leisure-entertainment program which is located in Guangzhou Shangjia plaza,over 2000 square meters,mainly built for the entertainment purpose for the urban white-collar.Dekang Club basic on the city landscape ,deconstruction the Tang dynasty blue and green landscape painting, make a respond for the appeal of the city life.

From the elevator to the reception, the walking stream of people is the designers breaking point,the twist and turning stream of people is just like the white cloud above the mountain.The triangle cutting of the wall is various,people will like this. From the reception to the canteen ,designers want to reflect a relationship from dynamic to still. In the two entrances of the reception hall and the entrance of the canteen, they put a crow-blue dry tree.Like the ancestors said: "Landscapes begins from the tree".Tree is the beginning, the leading elements.

挑战与技术实现：解构中国唐代青绿山水画及现代化技术实现过程。因墙体三角面的压纹不锈钢切割面拼装技术要求，在实施过程中，设计师在拼装上研发了一系列不同角度的五金构件。地面条纹石材也是现场作业实施的难点，无不体现了现场作业的匠心独到之处。

Challenge and presentation: Deconstruction the Chinese Tang dynasty blue-green landscape painting and the modern technique presentation. According to the technical requirement of installing all kinds of triangle press vein-like Stainless steel,their designers have developed various of hardwares.And the vein-like stone floor is also a difficult part to presentation.They do all this with their great effort. You will see that.

拥抱
ABRAZO

项目地点：台湾省台北市 /Location : Taibei, Taiwan
项目面积：66.5 平方米 /Area : 66.5 m²
公司名称：京玺国际股份有限公司 /Organization Name : Exclaim United Corp.
设 计 师：周谦如 /Designer : Joy Chou

IAI 设计优胜奖

周谦如 Joy Chou
中国台湾 Taiwan, China

2014 年　陶朱隐园艺术住宅招待会所规划设计（台湾 台北）；
2013 年　健鼎科技品牌整合设计及展示厅规划（湖北 仙桃）；
2011 年　LESS 杭洲思创服饰江南布衣集团（浙江 杭州）；
2010 年　Esquivel 好莱坞时尚明星品牌（北京 瑜舍酒店）；
2009 年　五叶神 烟草品牌年度广告设计及品牌整合（广东 深圳）；
2009 年　大百汇集团盐田企业总部设计（广东 深圳）；
2009 年　大百汇集团品牌整合设计（广东 深圳）；
2009 年　统一银座 Unimart 超市品牌整合设计（山东 济南）；
2008 年　Converse China R&D Center 中国 Converse 设计研发中心设计（广东 中山）。

2014　TAO ZHU YIN YUAN Reception Office Planning Design, Taibei,Taiwan;
2013　Tripod Technology Corporation Brand Design and Planning Design of Exhibition Hall,Hubei,Xiantao;
2011　LESS,Hangzhou,Zhejiang;
2010　Esquivel,Beijing;
2009　Wu Ye Shen Advertising Design and Planning Design, Shenzhen,Guangdong;
2009　DABAIHUI GROUP Headquaters Design,Shenzhen, Guangdong;
2009　DABAIHUI GROUP Brand Design,Shenzhen,Guangdong;
2009　Unimart Brand Design,Jinan,Shandong;
2008　Converse China R&D Center,Zhongshan,Guangdong.

有别于传统工业风格在生硬、冰冷线条下造成的空间距离感，以温暖的质素辅以人文的温度，借由视觉上铺陈温馨、亲近的场所个性，刻画所有的设计细节。融汇工业风格中的颓废元素，在红色砖墙的背景中，创造出令人印象深刻的感性情境。运用象征优雅、内敛的灰色，阐述简练时尚的咖啡色，以及具有现代意象的金属色调穿插其中。材料上以实木、铁件、布幔、仿旧金属网、复古砖为主，植入时尚、艺术的氛围。从灯光、材质、颜色综合营造出整体的氛围，汇入视觉艺术的墙面设计，仿旧金属网穿梭于全案的界面或吧台，复古砖亦包覆背景立面，同质性的素材与线性质感构成视觉线索，与场所的前后尺度张力呼应，借由串联艺术的美学张力，让空间自然流露出轻松、温馨的气息。

ABRAZO Different from the rigid and cold line of traditional industries that result in a distance sensation of space, the case featureswarm qualities supplemented with humanistic warmth to narrate the field personality of coziness and intimacy visually todepict all design details. Fused with the decant industry style, the red-brick background creates an impressive and yet profoundly sensational room by adopting the elegant .The use of grey color symbolizes reserved elegance; brown color expounds refined fashion while golden color with contemporary image is blended in between.The team applies wood,hardware, curtain, vintage metal gauze, and retro bricks to embed a carefree and relaxing ambient of fashion and art.The lighting, material and mixed colors create an overall ambient imported with the wall design of visual arts, where the vintage metal net passes through the interface and bars of the case. The retro-style bricks also cover the background fade while the homogeneous materials and liner qualities form visual clues to correspond with the front and rear scale intensity of the field. Such aesthetics intensity connecting arts provides a relaxing and cozy air naturally flowing in the space.

迷失酒吧
Lost Lounge Bar

项目地点：湖南省长沙市开福区万达国际广场开福金街内 /Location : Wanda Square, Kaifujin Street, Kaifu District, Changsha, Hunan
项目面积：476 平方米 /Area : 476 m²
公司名称：一本设计 /Organization Name : Eben design
设 计 师：李柱明 /Designer : Li Zhuming

IAI 设计优胜奖

迷失酒吧是一个以"鹿"为主题并全新定义的酒吧。全新的酒吧平面布局，相比从前有很多变化。吧台区域的全新组合方式，增加了人与人之间的互动性，顶棚的装置和吧台的组合方式与鹿角的交错形态相吻合。吧台采用亮面白色烤漆和透光玻璃，显得十分时尚、大气。顶棚装置同样采用亮面白色烤漆，视觉冲击力极强。鹿装置同样是亮面白色烤漆菱形构成体，时尚、简约、优雅，金色圆管组合强化了空间感。

LOST Lounge Bar was a deer as the theme and a new definition of a clear. Bar plane layout of new, compared to forward a lot of changes. The new combination of bar area, to increase the interaction between people, staggered form combination device and a bar of smallpox and antlers coincide. The bar with bright surface white paint and transparent glass, very stylish atmosphere. Ceiling device using the same bright white paint. A strong visual impact. Deer device is also bright face white paint rhombic structure, fashion and elegant simplicity. The combination of strengthening the sense of space golden tube.

李柱明　Li Zhuming
中国　China

2010 年毕业于广东轻工职业技术学院环境艺术系；
2014 任广州亦本良造室内设计有限公司设计总监。

2010 年　树影婆娑主题餐厅酒吧；
2013 年　长沙 LOST 迷失酒吧；
2014 年　长沙夜遇概念清吧；
2014 年　OPEN BAR；
2015 年　爱丽丝花园餐吧。

2010　Graduated from Department of Environment Art, Guangdong Industry Technical College；
2014　Design Director of Eben Design

2010　The Shadows of Trees Appear；
2013　LOST Lounge Bar；
2014　NIGHT EVENT；
2014　OPEN BAR；
2015　Alice Garden Restaurant。

音浪 KTV
Yin Lang KTV

项目地点：浙江省杭州市中大银泰城 /Location : Cuhk Yintai, Hangzhou, Zhejiang
项目面积：2 800 平方米 /Area : 2 800 m²
公司名称：杭州境致装饰设计有限公司 /Organization Name : Hangzhou Jingzhi Space Design Co., Ltd.
设 计 师：魏志学 /Designers : Wei Zhixue

IAI 设计优胜奖

魏志学　Wei Zhixue
中国　China

高级室内设计师，设计总监；
2008 年　创办境致空间设计事务所；
2004—2008 年　在杭州一所建筑设计事务所工作；
2003 年　毕业于沈阳工业大学室内设计专业。

Senior interior designer, Design director ;
In 2008　He founded in Jingzhi Space design firm ;
2004—2008　He worked in architectural design firm in Hangzhou.;
In 2003　He graduated from shenyang university of technology professional interior design.

音浪 KTV 中大银泰城店是此品牌于杭州的第三家连锁店。业主旨在将其打造成为杭城首屈一指的现代、时尚、个性化的量贩式 KTV。设计师也运用了很多的现代元素。原建筑为椭圆形的平面建筑结构。所以在整个设计中，大量运用椭圆形、曲线等不规则造型，与整个建筑风格相呼应。大厅的接待处，为一个圆形的大型前台，易形成视觉焦点。前台后通体 28 米长的弧形背景墙及等候长凳，都是由一块块 2 厘米厚的钢化玻璃拼成。再加以可变色的 LED 灯光烘托整个大厅的时尚氛围。

走道空间也是以曲线的手法呈现。渐变镂空的墙面发光灯箱，还有多层次的弧形造型，依附于整个设计的延伸方向，从小关系上加以变化。偶尔顶上的包厢及导视系统也闪耀着炫彩的灯光，以此种方式来展现它的与众不同。而在大大小小的包厢设计中，则采用了更加多元化的方式，或是将公共区域的设计手法运用至此，用多层次不规则的方式来打造一个迷幻空间。再或是将斑点、Hello Kitty 等女性青睐的元素搬入其中，突出别样的风情韵味。

Yin Lang KTV zhongda intime City store is the third chain store for this brand in Hangzhou.The owner aims to build a leading,modern, fashion, personalized volume KTV. We also use a lot of modern elements to this.
The original buildings for elliptic plane structure. So in the whole design, a large number of using this circular, curve, such as irregular modelling, and the whole building conditions. Reception hall, to a large circular desk, form the visual focus. Front desk connect body after 28 meters long arc setting wall and waiting for benches, are made of toughened glass Mosaic pieces two centimeters thick. Then can change color led lights foil of the fashion atmosphere of the hall.
Aisle space also appears in the technique of curve. Gradient the metope hollow out shine light boxes, and multi-level arc shape, attached to the extension of the overall design direction, since the childhood relationship to change them. Occasionally is out from the top of the box number and advertising system is glaring dazzle colour of light, in this way to show it's different.While in the large and small box design, adopted a more diversified way. Or use the public areas of the technique, multi-level irregular way to build a psychedelic space. Again or spots, Hello Kitty women favor of elements such as move, highlight the amorous feelings of different lasting appeal.

平面布置图

世辰财富会所
Shichen Fortune Club

项目地点：福建省长乐市广场路 13 号 /Location : No.13, Guangchang Road, Changle, Fujian
项目面积：233 平方米 /Area : 233 m²
公司名称：福建东道建筑装饰设计有限公司 /Organization Name : Fujian Dongdao Decoration Design Co., Ltd.
设 计 师：李川道 /Designers : Li Chuandao

IAI 设计优胜奖

李川道　Li Chuandao
中国 China

2013—2014 年　福建东道装饰设计有限公司总设计师；
2005—2012 年　福州东易装饰设计有限公司首席设计师；
2001—2004 年　福州大天装饰设计有限公司设计师。

2013—2014　Chief Designer of Fujian Dongdao Decoration Design Co., Ltd.;
2005—2012　Chief Designer of Fuzhou Dongyi Decoration Design Co., Ltd.;
2001—2004　Designer of Fuzhou Datian Decoration; Design Design Co., Ltd.

品茶由古至今都是一件十分雅致之事，不仅对茶品本身十分讲究，对于品茶的环境也要求很高。世辰财富会所以石材作为空间的主要用材，用黑、白、灰三种色彩营造了清雅的空间氛围。虽大量使用石材，空间却丝毫不显得枯燥无味，丰富变化的石材，以光面、毛面等多种形式展现，并结合在一起形成凹凸的立体感，为空间创造了多层次的视觉观感。入口的景墙由绿植打造，结合水景，自然风情弥漫空间。地面铺设青灰的复古砖，家具、装饰都选用黑色，带来沉稳的气息。

Tea-tasting has been regarded as elegant all the time, which not only has high requirements for the quality of tea but also for the tea-tasting environment. Shichen Fortune Club takes building stones as the main material and creates elegant space atmosphere with three colors – black, white and grey. Although lots of stones are used, the space does not appear to be dull. Changeful stone materials form a stereoscopic sensation with diverse ways such as glaze and rough surface, creating a multi-layered visual impression for the space. The landscape wall at the entrance is made of green plants, which presents a kind of natural style in the space together with the waterscape. The floor is paved with grey rustic tiles, and all furniture and decoration choose black, bringing a touch of calmness.

整个会所空间的围合感很强,利用屏风、景墙等作为隔断塑造小型的半封闭式空间,不论是落座于空间内,或是游走在走廊间都不容易互相干扰,并保留了空间的神秘感。长长的回廊,一面是粗糙的石板,一面是光滑的黑色木料,两种不同的色彩,不同的材质,形成有趣的对比,在射灯的光晕中投射出迷人的光影。在包厢与包厢间的间隙,辅以绿植、白色碎石等作为装饰,可谓是处处有景。包厢以透光的磨砂玻璃为墙,既保留了私密性,空间也能得到良好的采光。清新的环境,醇香的茶汤,约三五朋友聊尽世间百态。

The whole club space features strong enclosure. Screens, landscape walls etc are used as partitions to shape small semi-closed space. No matter where you are, sitting in the space or wandering through the corridor, you will not be interfered by others, thus keeping a sense of mystery of space. One side of the long corridor is rough slate and the other side is smooth black timber. Two different colors and different materials integrate with each other and present interesting contrast, sending out charming light and shadow in the halo of spotlights. There are green plants and white gravel as decorations between balconies, with beautiful scenes everywhere. Balconies are equipped with diaphanous ground glass as walls, not only retaining privacy but also ensuring good lighting. Fresh environment and mellow tea will allow you and your friends to enjoy talk.

国砚接待中心
State Banquet Reception Center

项目地点：台湾省高雄市苓雅区 /Location : Lingya Dist., Kaohsiung 802, Taiwan
项目面积：211.75 平方米 /Area : 211.75 m²
公司名称：橙田室内装修设计工程有限公司 /Organization Name : Chain 10 Urban Space Design
设 计 师：罗耕甫 /Designer : Kengfu Lo

IAI 设计优胜奖

罗耕甫 Luo Gengfu
中国 China

台湾橙田设计事务所，负责人；
上海思橙设计事务所，负责人、首席执行官。

福兴林宅（高雄，鸟松区）；
De Sede 瑞欧典藏家饰（高雄，四维路）。

Director of Chain 10 Urban Space Design ;
Director of Shanghai SIcheng Architecture ;
Design Firm.

Fuxing Lin residence;(Niao song district, Kaohsiung)
De sede Decoration.(Siwei street, kaohsiung)

本案尝试抽离古典的奢华，并且置入20世纪80年代的雅皮风格。空间中所有的功能需求皆以实用为主。矩形的开放空间划分出三种使用功能，在空间的区分上利用"功能"围塑出使用场所，使得空间的使用上更加宽敞与流畅，也因此让室内与外在环境的关系更加密切。

进入公共空间，大面积的茶镜顶棚串连全场，低反射度的镜面让空间中弥漫着古典气息，亦同时与雅皮风格的吧台产生互应。吧台选用消光面的玄武岩洞石包覆，石材的孔洞搭配凹凸错落的拼贴方式，在灯光的照射下呈现丰富多元的表情。长吧台在宽阔的空间中，发挥了动线引导的功能，给予使用者明确的行进方向，亦增加空间中的安定感。

家具搭配打破刻板的印象，以古典双人沙发和雅皮风格扶手椅的出现转化与人之间的互动关系，为沙发区带来戏剧性的冲击力，并形成风格上的强烈对比，流露出了整体空间欲阐述的讯息，一切的矛盾与冲突在此得到了充分的诠释。

The luxury of classic is pulled away in an attempt, and yuppie style of the 80s is added in the project. All the functions and needs of the space are based on Pragmatism.Rectangular open space is divided into three kinds of function. The division of the space is based on the function. It makes the use of space to be more spacious and smooth, therefore, the indoor and outdoor's connection become much closer.

The large tawny mirror arranged on the ceiling with low specular reflection in the public space makes the space filled with classic atmosphere.It is corresponding with the yuppie-style bar. The dull surface travertine of basalt is selected to cover the bar. The scattered collage and stone holes present the rich and divers expression under light irradiation. The long bar in a wild space plays the function of moving line to indicate the users a clear direction and enhance the sense of stability.

The arrangement of furniture breaks the stereotype.The classic double sofa and yuppie armchair transform the interaction with people and add the dramatic impact to the sofa area.It becomes a strong contrast of style to reveal the message of the overall space.There has been an adequate interpretation of contradiction and conflict.

环上大影视园区大师工作室
Master Film Studio in the Rim of Shanghai University

项目地点：上海市闸北区万荣路700号 /Location : 700 Wanrong Road, Zhabei District, Shanghai
项目面积：974平方米 /Area : 974 m²
公司名称：华东建筑设计研究院有限公司 /Organization Name : East China Architectural Design & Research Institute
设 计 师：徐访皖 /Designer : Xu Fangwan

IAI 设计优胜奖

徐访皖　Xu Fangwan
中国　China

2006年3月　高级室内设计师；
2005年6月　高级工程师；
2003年1月　室内建筑设计师；
1994年至今　华东建筑设计研究院有限公司；
1994年　毕业于上海同济大学。

2006　Senior Interior Designer;
2005　Senior Engineer;
2003　Interior Architecture Designer;
Since 1994　Work at East China Architectural Design & Research Institute;
1994　Graduate from Tongji University.

新落成的环上大影视园区大师工作室，是由老厂房改造的工作室，吸引了国际知名电影制作人和一批青年新锐导演落户。设计中充分挖掘了老厂房中的特色，加以保留和运用。室内设计选用的饰材，既体现了工业化的元素，又表达了绿色和环保的设计理念。
入口区的设计，让人产生户外影院的联想。吊顶采用弧形的软膜，阵列排布，仿佛是天幕电影，配合声光电的变化，欲以先声夺人。

The newly built Master Film Studio in the Rim of Shanghai University has attracted world-known international film makers and young new directors. The design has tapped the characteristics of old factory to preserve and make use of it.The decoration materials of the interior design show the industrial elements as well as the idea of green and environmental protection.
The entrance design makes a picture of outdoor cinema. The usage of arrayed arc film and alternation of sound and light create an unreal and shocking feeling.

一层新锐导演区设置两个放映室，一个封闭，一个开放。开放区亦成为大家交流与休息的空间。位于后区的服务吧台为各种活动提供便利。二层为大师工作区域，厂房的高空间、高侧光带来独特的视觉效果，与走道互成角度的隔墙不仅使入口处看上去较为私密，也丰富了空间效果，巨幅的电影手工趣味画，让人保持和呼唤创意的热情。

Floor 1 the Area of Young New Directors have 2 screening rooms, one open and the other closed. The opening area enhances the relaxation and communication of audience, and the service bar located in rear area provides convenience for various activities. The second floor is the Master Studio. The high storey height and high side light have brought unique visual effects; the partition wall does not only make the entrance secret, but also enrich spatial effects; and the hanging of huge hand-drawn fun pictures stimulate the creation of directors.

龙圣汇会所
Long Shenghui Club

项目地点：北京回龙观区域 /Location : Huilongguan District, Beijing
项目面积：2 800 平方米 /Area : 2 800 m²
公司名称：杜玛（香港）国际有限公司 /Organization Name : DOMO Nature Co.,Limited.
设 计 师：赖亚楠 /Designer : Lai Yanan

IAI 设计优胜奖

赖亚楠 Lai Yanan
中国 China

中央工艺美术学院环境艺术设计系本科毕业，建筑学院研究生毕业；
从事景观、建筑、室内及家具和陈设艺术品的设计；
2012 年 米兰国际家具展——"坐下来"中国当代坐具设计展；
2011 年 北京国际设计周"享自.东方"——中国原创设计展；
2004 年 在法国巴黎创立了原创设计品牌 DOMO nature ；
1998 年 率先在国内提出，倡导并积极实践"一体化的系统整合性设计理念"。

Bachelor degree in environment art design of Central Academy of Art and Design,Graduate degree in school of architecture.Engaged in landscape architecture interior and furniture and display art design;
In 2012 Milan International Furniture Exhibition—"Sit Down", China Contemporary Design Exhibition;
In 2011 Beijing Design Week. "Enjoy.China" China Original Design Exhibition;
In 2004 founded the original design brand DOMO nature in Paris,France;
In 1998 the first to put forward the concept of the integration of the system design.

龙圣汇会所位于北京回龙观区域的黄金地段，共 3 层，面积为 2800 平方米，是一个定位为主题书房的商务社交场所。空间设计旨在打造一个安宁的场所、有意义的豪华，清新的人性，表达一定意义的风格与内涵，期待一致性的礼遇。本项目是从建筑到室内到家具陈设品和技术呈现的一个完整性的，以一体化的思路和形式来完成的综合性商业空间。其内部功能集商务宴请、洽谈、住宿等为一体。空间很好地体现了视觉的张力。用素雅、含蓄的手法展现了一种清新的格调。

Long Sheng Hui Club, seated in the prime location in Hui Long Guan, Beijng, is a three-storey building and covers an area of 2,800 square meters. It is designed to be a study-room themed club offering services for business and social activities. The space design aims to provide every customer with a distinguished experience in an ease, luxury and people-friendly environment highlighting on its artistic style and content. From indoors to outdoors, the club is a complete package for hosting commercial activities integrating building structures, furnishing and technical support. The Club is multi-functioned combining features for hosting business banquet, meetings and for accommodations. The design and decoration, connotation in method, delivers an elegant and refreshing style and provides clients with an excellent visual feast.

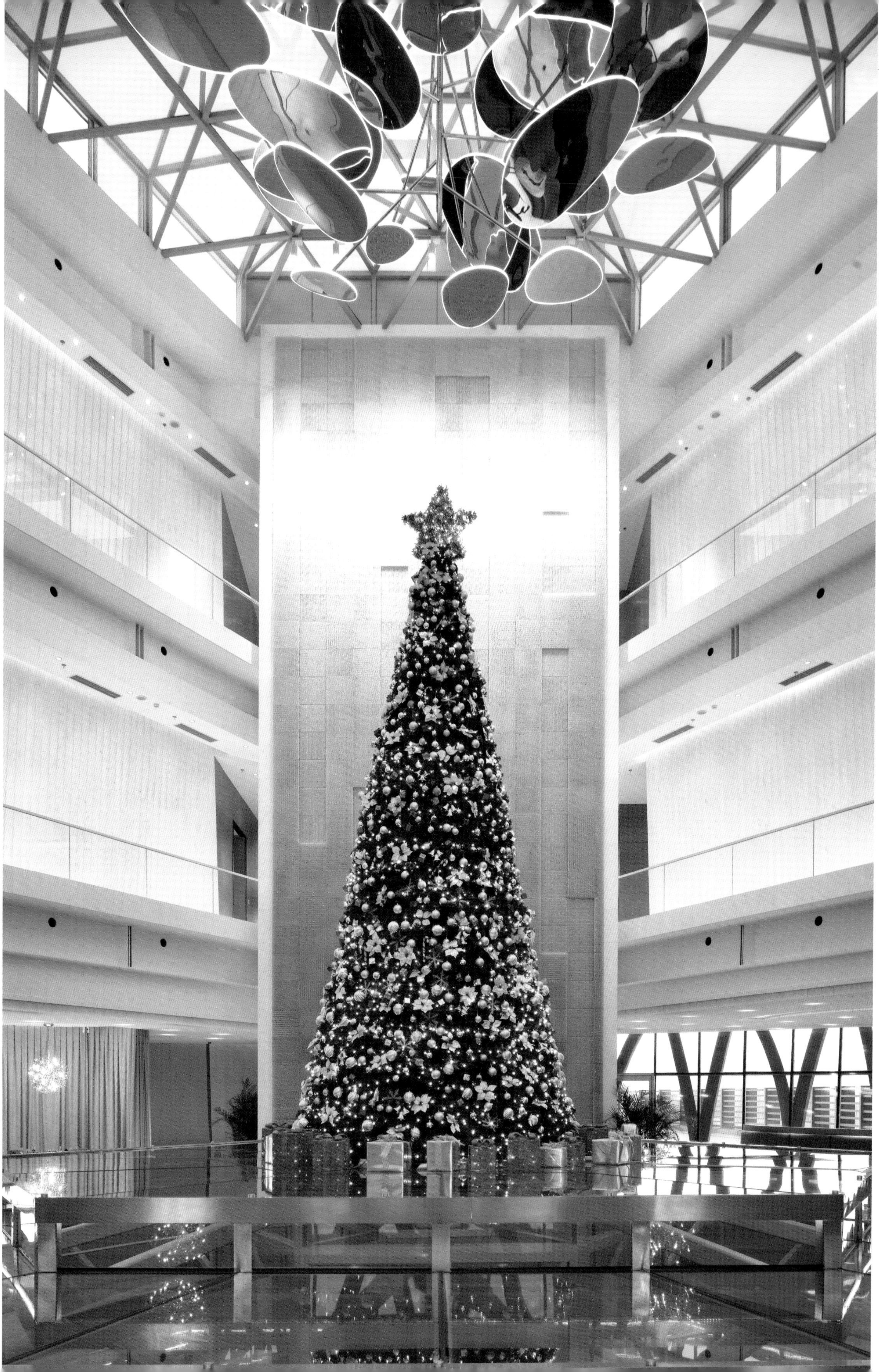

Culture and Exhibition Space
文化、展示空间

Music Classroom	96
Steps	98
State Grid JiangSu Electric Power Experience Centre	102
China Optical Science and Technology Museum	104
ECLAT (1F)	106
ECLAT (8F)	108
Qiantang Residence Show Flats B , Hangzhou	110
Mimicry of Light	112
One China One World	116
The Fashion Architecture	118

96	音乐研习中心
98	拾阶
102	国家电网·江苏省电力体验馆
104	中国光学科学技术馆
106	ECLAT(1F)
108	ECLAT(8F)
110	杭州钱塘会馆样板间 B
112	光景拟态
116	一瓷一世界
118	纵横天厦会所

音乐研习中心
Music Classroom

项目地点：台湾省台北市板桥区中正路 41 号 /Location : No.41, Zhongzheng Road, Banqiao District, Taibei, Taiwan
项目面积：170 平方米 /Area : 170 m²
公司名称：金湛室内设计有限公司 /Organization Name : Goldesign Studio
设 计 师：凌志谟 /Designer : Ling Chih-mo

IAI 设计优胜奖

以原始为出发点，此空间希望呈现出明亮和轻快的格调，以材料原始的样貌，加上镀锌钢板烤漆、钻泥板的纤维、人工草皮来表现自然质朴简约的风格。外观以纯净的白色做为整体建筑的基调，点缀上些许象征自然的木材，以及线条简单的烤漆钢板，呈现出一种自然质朴的风格。入口处采用大面积简单穿透的落地玻璃，做为室内与室外的分隔，让人有种空间宽敞延伸的感觉。没有过多的装饰，简单的弧形线条表现在外观及室内，没有一般教室的刻板印象，取而代之的是让人可以放松享受音乐的场所。

Primitive is the origin of the design. Keeping theoriginal look of the materials: uncoated galvanized steel, Diacrete, plus artificial turf, the space aims to present a bright, lively yet natural and primitive style.White is adopted as the general base color for the exterior, decorated with a wooden and a pre-painted galvanized steel fascia.A large full-height glass wall is used for the entrance to separate the exterior and interior,while extending the space.Without excessive decorations but simple curvy lines, the space broke free from a stereotypical classroom, and is one in which people can actually relax and enjoy the music.

凌志谟 Ling Chih-Mo
中国台湾 Taiwan, China

2007 年创立金湛空间设计研究室 Goldesign Studio，致力于透过业主的内涵与现代风格反映出生活的空间时尚品位和文化层次，作品崇尚简约，线条比例充满美感。

透过创意延伸居住者的生活精神与内心期望，跳脱既定符号的框架与设限，逐一引导居住者从情绪酝酿至心境内化，透过界限之间的转折层次，触动个人深藏的记忆想念，寻求生活中种种美好的可能。

Ling Chih-mo established Jin Zhan Interior Design Co., Ltd. In 2007. He wants to show the taste of life space and culture according to the owner's characteristic and modern style. The feature of his work is simple, graceful lines.

Through the creative, extend the residents' spirit of the life and desire, jump out of limits and restrictions, guide habitant from brewing mood to internalization. By way of cross of different limits, touch people's deep memories, visit all sorts of beautiful in the life .

Culture and Exhibition Space 文化、展示空间

拾阶
Steps

项目地点：台湾省台北市忠孝东路 /Location : Zhongxiao East Road, Taibei, Taiwan
项目面积：190 平方米 /Area : 190 m²
公司名称：欣琦翊设计有限公司 /Organization Name : C.H.I. Design
设 计 师：何欣，林真琦 /Designer : Ho Hsin, Lin Chen-chi

IAI 最佳设计大奖

C.H.I. Design Studio

何欣 Ho Hsin   林真琦 Lin Chen-Chi
中国台湾 Taiwan, China

2008 年　欣琦翊设计有限公司项目设计主持；
2008 年　中原大学室内设计系 兼任讲师；
2003—2008 年　太一国际设计 项目设计。

Since 2008　Director of project design of CHI Design Studio;
Since 2008　Adjunct Lecturer at Department of Interior Design of Chung Yuan Christian University;
2003-2008　Project designer of Taiyi International Design Company .

本案原为旧式挑高空间且有区分上、下层的住宅。在空间的配置上，有高于入口处约 150 厘米的级高落差，为能够吸引来往人流目光，进而登上阶梯走入店内，于是创造出一个大型的黑色入口以斜度引导的方式，还有利用黑色烤漆玻璃的材质，经由不同视角随之而生的折射变化，建构视线焦点的交汇。
由于空间为狭长型相连的格局条件，中央如腰带般缩拢的过道区域，可能构成消费者于展示空间游走时的退却感。于是通过材质的变化将空间串连，并将贩售鞋品处依代理品牌的不同逐一划分，产生区域在各自氛围外仍具相互连结关系，再经由点光源的焦点式照明营造戏剧效果，在转换、跳脱之间烘托出行进走动的好奇感。在材质的选择上，整体空间以冷色调的灰色水泥板贯穿全场，希望以单一材质去贯穿空间来呈现低碳创意的环保概念。

This was an old residential space which had high ceiling and interlayer. Regarding the layout, there is a 150cm fall on the entrance so designers create a big black stair entrance with light steps as visual focus. Due to it has to catch the attention of people who pass by, and furthermore to let people take the stairs up into the store. designers need to construct the confluence of vision and attention, they use the material of black glass as the wall of entrance so that it can be seen deferent refraction following each light step up to the store.

In this space, there is a narrow path in the middle area, which could make customers stop their steps when shopping in this space. So, designers connect the space by changing materials and planning the areas for each shoe brand. It makes each deferent area has their own style but still connect with each other, and feature spotlight creating dramatic feeling when stepping through deferent areas.Regarding materials, to reduce consumption of resources by choosing concrete slab with cool gray to penetrate all space. No more decorating materials showing less-is-more and eco-friendly.

通过入口处的阶梯概念体现空间内部规划配置的布局，包含层次堆栈的透视墙面、商品呈现的展示柜体，皆由"阶梯"作为主题一以贯之。阶梯踏面和鞋品展示柜体切削出的60°斜踢脚收边，则源自其品牌名称及商标识别，以商标的正三角形各内角角度为出发点，呼应品牌理念的同时，并考量拿取商品时脚边站立的距离，使顾客与展示柜体之间有最适宜的观赏关系，将品牌形象与空间设计规划相互连结。

From the entrance to the space inside, all take "steps" as the major concept, including the perspective wall with layer stacked and display cabinet. there is a 60 degree angle skirt along all the cabinet because of the client's logo which is regular triangle. One hand designers response to the client's brand, on the other hand, 60 degree as skirting line can make perfect distance for shoppers who want to choose products on the display cabinet. So that successfully connect with client's image and display design.

Culture and Exhibition Space 文化、展示空间

国家电网·江苏省电力体验馆
State Grid Jiangsu Electric Power Experience Centre

项目地点：江苏省南京市建邺区奥体大街 /Location : Aoti Street, Jianye District, Nanjing, Jiangsu
项目面积：2 252 平方米 /Area : 2 252 m²
公司名称：上海尚珂展示设计工程有限公司 /Organization Name : Shanghai Thinker Exhibition Design Engineering Co.,Ltd.

IAI 设计优胜奖

上海尚珂展示设计工程有限公司
Shanghai Thinker Exhibition Design Engineering Co.,Ltd.
中国 China

尚珂展示是一家集新型展示空间设计、展品展项设计、工程实施服务为一体的展示空间专业设计机构。忠于"思行合一、臻于至善"的企业信条，用设计的力量打开商业世界的想象力。整合科技、艺术与设计，以新策略、新模式、新媒体为客户创造商业模式新体验。服务领域涵盖了商业场馆类、企业展示类、大型办公空间类等商业展示空间和博物馆、科技馆等公共文化空间。

"Thinker Exhibition Design"is a professional design company that offers new exhibition space,display elements design,construction services ,etc. "Thinker Exhibition Design"adheres to the idea"thought and behaviours in one, the pursuit of refinement ",and tries to open a new business world.Combining technology,art and design, creating a new commercial pattern(new strategy,new pattern and new media) for clients "Thinker Exhibition Design"focuses on the design of commercial space, exhibition space and grand office space, public and cultural space such as museum and technology and science museum,etc.

江苏省电力体验馆，展示面积为2252平方米，净高为7米，将采用先进的展示技术和手段，打造电力行业同类展馆中唯一一个以客户体验为中心的电力展馆。本案在布局上充分结合建筑特点，空间与展品展项结合紧密，形式与展示内容力求统一。在实施上执行场外加工、场内安装的总体思路，大量运用GRG、铝板等成品定制材料，既保证了材料品质，也确保了正常的施工周期。

JiangSu province electric power experience pavilion,the display area is 2252m², and it is about 7meters in height .It will use the advanced display technique to build the only one electric power exhibition based on customer experience in similar pavilion .The layout pursues close combination of spaces and items on display, as well as the unity of form and content. The guideline of implementation is processing out of the pavilion and assembling in it, meanwhile a lot of finished and customized materials such as GRG, aluminum plates are used to assure both material quality and construction period. After put into operation, the pavilion has been visited by many people from power industries and other walks of life, and is highly praised by them.

Culture and Exhibition Space 文化、展示空间

中国光学科学技术馆
China Optical Science and Technology Museum

项目地点：吉林省长春市净月开发区永顺路1666号 /Location : No.1666,Yongshun Road, Jingyue Economic and Development Zone, Changchun, Jilin
项目面积：约3 000平方米 /Area : about 3 000 m²
公司名称：上海尚珂展示设计工程有限公司 /Organization Name : Shanghai Thinker Exhibition Design Engineering Co.,Ltd.

IAI 设计优胜奖

设计师在设计色彩上，分别以"赤、橙、黄、绿、青、蓝、紫"的七色光作为各个展厅的基础色彩；根据每个展示空间的主题及内容，提炼与升华，利用麦穗、山峰、万花筒、交织的网等精彩的设计元素，让展示空间的内容与形式达到完美的和谐统一。

The designer uses red, orange, yellow, green, indigo, blue, purple as basic colors of each exhibition hall, and makes decoration all related with light on door pockets, ground, ceiling lighting and walls to create a brilliant optical world that surrounds visitors and brings them wonders and joy. Each changing exhibition space is clearly and smoothly connected by moving lines, and symbols of elements are extracted by content extension and symbolic significance.

ECLAT (1F)
ECLAT (1F)

项目地点：台湾省新北市五股区五权路 28 号 1 楼 /Location : 1F, No.28, Wuquan Road, Wugu District, Xinbei, Taiwan
项目面积：700 平方米 /Area : 700 m²
公司名称：品昕空间计划有限公司 /Organization Name : CCPLUS Design Co., Ltd.
设 计 师：马静自 /Designer : Ching-Chih Ma

IAI 设计优胜奖

品昕空间计划有限公司

马静自 Ma Ching-Chih
中国台湾 Taiwan, China

美国 GC DESIGN 集团联合创办人；
全球设计师联合协会（GDADA）美国常务理事；
台湾教育部认证大学专业讲师资格；
品昕空间建筑计画事务所，创办人，设计总监；
极致空间设计，设计总监；
汇侨设计（RICH HONOUR），精品品牌设计部；
汇庭宽室内设计，设计师；
台湾大叶大学空间设计系建筑组，室内设计组；
2004 至今现任台湾教育部大学讲师。

Co-Founder of GC DESIGN GROUP (USA)
Global Deco America Design Association;
The Ministry of Education Professional Certification;
University Lecturer;
Design Director and Founder of CCPLUS Design Co., Ltd.;
Design Director of Jizhi Space Design;
Brand Design Department of RICH HONOUR;
Designer of Huiting Interior Design;
Dayeh University, Taiwan;
Since 2004 Architecture and Interior Design Instructor.

这个展示厅需要借由多方的形式，来展示不同风格的服装，并将多样的设计产品与其功能展现给顾客。展示设计的主要概念，是自然的风格结合自由的理念。首先，借由庞大弯曲的不锈钢管，穿梭于空间当中，创造一个自然循环的方式，象征着生命的循环，使空间形成一脉的串连。其次，矩形木柜体，吊挂在圆形轨道之下，顾客可以推动木柜，使物品随着轨道循环移动，借由人们与道具的互动，活跃了空间的氛围。最后，七彩玻璃生成的立方体，则代表着多彩的世界及光的纹理。

This showroom needs versatile display to address variety of the wears in order to show to their clients both design and multifunction products. Within the total meaning of display design is focusing on natural and free style. First of all, in the front, the giant curvy stainless steel pipes creates a natural circulation style represents circle of life, Second, wooden boxes hang on a circle rail which represents recycle idea. People could push the boxes and make items move around the rail. Third, colorful glass cubics represent colorful world and light textures.

ECLAT (8F)
ECLAT (8F)

项目地点：台湾省新北市五股区五权路28号8楼 /Location : 8F,No.28, Wuquan Road, Wugu District, Xinbei, Taiwan
项目面积：700平方米 /Area : 700 m²
公司名称：品昕空间计划有限公司 /Organization Name : CCPLUS Design Co., Ltd.
设 计 师：马静自 /Designer : Ma Ching-Chih

IAI 设计优胜奖

甲方是以纺织为主要生产重心的世界级大型公司，其主要产品为纺织品及其所设计的服装。这次的设计方案主要是希望将旧有的传统文化，以一种创新手法来赋予其新的生命，使用当地的材料，将纺织的元素置入空间设计中，除了让传统的纺织业呈现新的样貌外，也将地性的精神传承下来。对于纺织而言相当重要的水资源，以及同时为纺织及服装所注重的色彩元素与织线元素，皆作为空间之间的连结，将传统的精神萦绕于空间当中，保留文化，并创造新的态度。

This major world-class company is focusing on manufacture of textile and clothing as well as designing sportswear for men and women. In order to recreate for this company with a new image and provide a variety of possibility for their international customers and designers a perfect purchase experience with a spatial integrity showcase of the company's products. Due to this company with a highly requirement of designing and colour demanding, therefore, to redefine the space between interior and fashion design as a critical issue here. Transforming the form of water fall cubes as considered the most important concept and feng-shui resource of the company since when this company was established.

杭州钱塘会馆样板间 B
Qiantang Residence Show Flats B, Hangzhou

项目地点：浙江省杭州市江干区钱江新城剧院路 99 号 /Location : No.99,Theater Road, Qianjiang New City, Jiangganzhou, Zhejiang
项目面积：约 400 平方米 /Area : about 400 m²
公司名称：艺玺空间设计（上海）有限公司 /Organization Name : IVAN C. Design Limited
设 计 师：郑仕梁 /Designers : Ivan Cheng

IAI 设计优胜奖

郑仕梁 Ivan Cheng
中国香港 Hong Kong, China

毕业于香港理工大学室内设计系；
香港演艺学院 舞台布景及服装设计系；
曾就读于香港珠海大学室内建筑系；
香港 IVANC.DESIGN LIMITED 总经理兼设计总监；
郑仕梁室内设计（上海）有限公司总经理兼设计总监。

Graduate from Department of Interior Design of The Hong Kong Polytechnic University;
Graduate from Department of Scenery and Costume Design of The Hong Kong Academy for Performing Arts;
Study at Department of Interior Architecture OF Chu Hai College of Higher Education;
Design director &General manager of IVANC. Design Limited (Hong Kong);
Design director &General manager of IVANC. Design Limited (Shanghai).

杭州钱塘会馆样板间 B 以新中式手法为主，以"水"的内敛为主题，具有西方的外表和东方的内在，平静、简洁、自然的空间。享受的是超越物质的空间，得到心灵的宁静。呈现在主人眼中的不再是具象的线条与色彩，而是水墨般的柔软线条与自然色泽，融汇出的却是身处湖中或天空中的美妙意境。

Qiantang Residence Show Flats B mainly adopts the new Chinese-style technology, with the introversion of the "water" as its theme, possessing Western appearance and Oriental inherence, as well as a calm, simple and natural space. What you enjoy is a space beyond the material, and what you get is the peace of mind. What in the master's eyes are no longer figurative lines and colors, but ink-like soft lines and natural colors, rendering a wonderful artistic concept of living in the lake or sky.

光景拟态
Mimicry of Light

项目地点：台湾省桃园市 /Location : Taoyuan, Taiwan
项目面积：160平方米 /Area : 160 m²
公司名称：彩韵室内设计有限公司 /Organization Name : Cai-In Interior Design Co., Ltd.
设 计 师：吴金凤，范志圣 /Designer : Chin-feng Wu, Chih-sheng Fan

IAI 设计优胜奖

吴金凤 Chin-feng Wu
中国台湾 Taiwan, China

彩韵室内设计有限公司，设计总监；
京采室内装修工程有限公司，总经理；
彩琚国际室内设计有限公司，执行总监；
内政部建筑物室内设计师；
内政部建筑物工程管理师；
SICK-HOUSE 病态住宅诊断士；
万能技术大学业外讲师。

Design Director of Cai-In Interior Design Co., Ltd;
General Manager of JT Design Company;
Executive Director of Caiju International Interior Design Company;
Interior Designer of Ministry of the Interior;
Project Manager of Ministry of the Interior;
Diagnostician Of SICK -HOUSE;
Guest Lecturer of Vanung University .

公共空间
木料板材的拼贴起伏，将客厅主墙布置为立体的图层，其上，两幅油画的朦胧笔触彷彿垂落于深色系的沙发与木桌，灰蓝、釉青、奶白，游离于抽象的色彩域度，塑制为靠垫、花器。装饰物件的颜色及重点家具的材质、样式并非戏剧化的元素，空间感宛如一条静谧之河，轻轻泛流，濡湿了客、餐厅的半隔间石材，将纹理磨砺出岁月的意味，工业风糅合中式的造型灯具，成为自然素材之间的聚焦，透过半隔间的"留白"，则可框取客厅墙景。

Public space
Wooden planks are pieced together like a jig puzzle. The main wall of the living room was presented as a contoured picture.Obscure brush stokes of the two oil painting seem to extend directly to the dark sofas and wooden tables, with livid, enamel green , and white milk colors departing from the abstract and entering reality to mold the sofa cushions and flower vases. The color of the decor and materials of key furnishings and styling has created sober elements where space is likened to a serene river flowing slowly to moisten the partial stone partitions of the living and dinning rooms. The patterns assumed a weathered appearance, supplemented by industrial style lighting with traditional Chinese elements, has become the focal confluence of natural materials, with the blank in the partial forming a frame around the living room wall.

私密空间

相较于公共空间宁静中富含生机的感受，主卧室突显的是前卫个性的面貌：斑驳墙面、工业风灯具、置于相框一角的黑白地景摄影，似要隐述现代城市的荒漠寓言，木质地板与边几、腰柜则适时平衡了清冷调性。另两间卧室亦采取简约的工业风格，灯具与墙面共构线条肌理，边几的装饰物件则呼应客厅桌面的陈设，以玻璃、彩色水晶与金属器皿为主，流露亲和之意。

Private space

In contrast to the serene yet vitalizing atmosphere provided by the public spaces ,the master bedroom emphasizes a maverick setting with patched walls, industrial-style lighting ,and a black-and-white landscape photograph placed at a corner of the picture frame, illustrating the metaphor of the urban desert. Wooden floors, side tables, and waist-high cabinets balance the cold and distant character of this setting. The other two bedrooms have also adopted a simple industrial style with lighting and wall surfaces generating cohesive contours.Decor on the side tables correspond to the living room table layout of glass, colored crystals, and metal utensils to provide warmth and intimacy.

一瓷一世界
One China One World

项目地点：广东省佛山市顺德区 /Location : Shunde, District, Foshan, Guangdong
项目面积：2 000 平方米 /Area : 2 000 m²
公司名称：香港亮道设计顾问有限公司 /Organization Name : Hong Kong Liangdao Design Consulting Co, Ltd.
设 计 师：关升亮 /Designer : Guan Shengliang (Ansun Guan)

IAI 设计优胜奖

关升亮　Ansun Guan
中国　China

香港亮道设计顾问有限公司现任董事、首席设计；
2000 年　开创个人事业，合伙创立高天创意设计中心；
致力于设计价值的挖掘和设计品质的完善；
从事室内专业设计 19 年。

The chief design director bright way design consultant co., LTD. in Hong Kong ;
Start his career in 2000, co-founder high creative design cente ;
Dedicated to the design value of thTe mining and improve the design quality ;
Professional in interior design for 19 years.

"一瓷一世界"是顺德当代一位年轻的女收藏家的私人艺术馆，以收藏展示古瓷，当代景德镇新瓷为主。馆主热爱传统文化，以传播当代东方文化为己任，勇气可嘉。艺术馆地处顺德桂畔河，是当地有名的文化区，如何让古瓷、新瓷与当代展示空间融合，并展示其东方魅力？设计师通过沟通，最后以"桂畔河畔书苑"为设计概念。古书苑是古代学者传播知识文化、讲学论道的地方。设计师借鉴古书苑的元素和底蕴营造出当代东方氛围的展示空间。"一瓷一世界"承载着一位年轻收藏家的梦想，也是设计师对当代东方设计语言的又一次尝试和探索。

"A porcelain One World" is a private art gallery of contemporary young woman Shunde collectors to showcase collections of porcelain, contemporary Jingdezhen porcelain new main. Main hall love traditional culture, to disseminate contemporary oriental culture of responsibility, courage. Museum of Art is located in Shunde River Kwai River, is famous cultural district, how to make porcelain new porcelain fused with contemporary exhibition space and show its oriental charm? Finally, the designer of "River Kwai River Bookstore" for the design concept. Bookstore ancient scholar of ancient cultural dissemination of knowledge, lecturing on the road in places. Designers draw Shuyuan ancient heritage elements and create a contemporary exhibition space oriental atmosphere. "A porcelain one world" carries a young collector's dream, but also the designer of contemporary oriental languages and again to try and explore.

纵横天厦会所
The Fashion Architecture

项目地点：台湾省台北市大安区四维路 168 号 /Location : No.168, Siwei Road, Da An District, Taipei, Taiwan
项目面积：2 000 平方米 /Area : 2 000 m²
公司名称：大间空间设计有限公司 /Organization Name : DAJ Interior Design Co., Ltd.
设 计 师：江俊浩 /Designers : Chiang Chun-hao

IAI 设计优胜奖

江俊浩 Chiang Chun-hao
中国台湾 Taiwan, China

2010 年　华夏技术学院室内设计系 兼任讲师；
2008 年　英国 University of Brighton 建筑设计硕士；
2008 年　DAJ 大间空间设计有限公司创立；
1996—2005 年　M+W 建筑空间研究所设计部经理。

2010　Adjunct Lecturer of Department of Interior Design of Hwa Hsia Institute of Technology；
2008　Master of Architecture of Brighton University;
2008　Establish DAJ Interior Design Co. Ltd;
1996-2005　Manager of Department of Design of M+W Group.

会所以时尚美学为设计概念，搭配流动的线条创造出自由混搭、不受拘束的空间设计创意以流动的飞扬、几何的稳定形塑出整体空间的基调，借此创造出强烈的特殊美感体验。空间构图上分为"流动"与"静定"。流动、飞扬的张力及静定、柔软的冥想凝思。简单中仍富有变化和韵律，将空间的生命力转化为内敛的境随心转。

The estate clubhouse - Fashion aesthetic design concept with the flowing lines create the unfettered freedom to mix and match creative space design , with flowing flying geometric stability shaping the overall tone of the main space , thereby creating a strong special aesthetic experience.Spatially patterned into " flow " and "calm" flow - flying within soft meditation meditation. Simple changes and still full of rhythm , the vitality of the space into a restrained environment mind transforms circumstances .

入口大厅大量留白犹如画布，让巨形雕塑成为视觉主角，招待会所则以柔软、律动包覆出空间美学。停车场呼应整个建筑外观并延伸至室内空间，转接到地下停车场空间，依续视觉流动的形式引导着回家的主人。

Entrance hall like a lot of blank canvas to make a giant sculpture sway become the visual concept of the protagonist , the reception of the soft places ,covering a space of aesthetic rhythm.
Parking echoed throughout the exterior of the building extends to the interior space, referred to the underground parking space according to the type of visual flow continued to guide the home owner.

Office space
办公空间

The 1122 Space Solidification	122
"Interaction": BWM Office	126
SanYin Group Shanghai Office	130
CONVERSE R&D CENTER	132
Ding Tian Decoration Company.	134
Samlee Office	136
Creative Origin	138
Sunlight in the Wood	140
Simplicity · Headquater of Shanghai Camerich	142
The Office Space of MAIDAO Realestate	144
Samson Wong Design Group Office	146
VIPABC	150

122	凝固的1122空间
126	"互融":博炜曼办公空间
130	三银集团上海总部办公室
132	匡威研发中心
134	鼎天装饰办公空间
136	Samlee办公室
138	初·原色办公空间
140	木制光影
142	简谧·上海锐驰总部
144	麦道置业办公空间
146	康华室内设计办公室
150	VIPABC

Office Space 办公空间

凝固的 1122 空间
The 1122 Space Solidification

项目地点：上海市 /Location : Shanghai
项目面积：135 平方米 /Area : 135 m²
公司名称：上海亿端室内设计有限公司 /Organization Name : Yiduan Shanghai Interior Design
设 计 师：徐旭俊 /Designer : Xu Xujun

IAI 最佳设计大奖

徐旭俊 Xu Xujun
中国 China

上海亿端室内设计有限公司 首席执行创意总监；
国际注册高级室内设计师；
中国室内设计师协会专业会员；
2012-2013 年 中国室内设计师年度封面人物。

Chief Executive Creative Director of Shanghai Yiduan Interior Design Co.,Ltd.;
Senior Interior Designer of International Accreditation and Registration Institute;
Professional Member of China Interior Designers Association;
2012-2013 The Cover Person of China Interior Designers Association.

凝固的1122空间是徐旭俊的设计工作室，数字是这间工作室的门牌号码，而"凝固的空间"则寓意着室内设计与建筑、雕塑的融洽关系。设计师既做自己的甲方又做乙方，办公区、会议室、接待区、展示区及简易的休息间等办公功能需一应俱全，对空间的布局有着苛刻的要求，更要在专业层面上别具一格，营造出宁静而质朴的氛围。

The Solidification of 1122 Space is the design studio of Xu Xujun.The digital part is the studio house number,and the " Solidification of 1122 Space "means a harmonious relationship among interior design, architecture and sculpture. The designer can be their own party A and party B, office, meeting room, reception area,display area and simple rest room, exhibition area, simple rest area and other functions. The layout of the space has a demanding, more on the professional level and have a style of one's own, creating a quiet atmosphere.

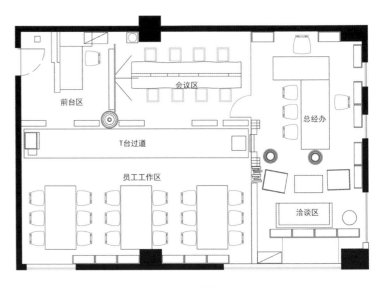

平面布置图

设计师运用灰质砖特有的粗砺感使空间回归自然,以书本形式的建筑材料塑造具有雕塑感的艺术空间,取得了直通心灵的效果。在整个空间色彩上以灰白为基调,而刷了木蜡油的老木桌面,以温暖的质感平衡了空间色彩的灰白中性基调,完成了温暖与冰冷的对话。地毯般木板走廊,红绿色油漆斑驳,被架空抬高后打造出空间的层次感,放置枯木的黑色陶罐,缠绕了几圈麻绳,既在色彩上有所呼应,又与前台顶上由麻绳交错编织而成的"凉棚"达成材质上的呼应,空间虚中有实、实中有虚。强烈反差的对比,使整个空间充满了另类诗一般的意境美感,在办公项目中达到了独具匠心的完美创意。

The use of coarse gray brick unique sense of space make a return to nature, in book formbuilding material shaping sculptural art space to create the soul through the effect. In the whole space color with gray tone, brush wood wax old wooden table, retains thenatural texture of wood, to warm the texture gray neutral tone balance of color space,also contains a naturalistic rustic materials, finished the warm and cold dialogue in the same space carpet wooden corridor, red green mottled paint, are overhead elevation to createspace level sense, placing dead wood black pottery, wrapping laps hemp rope, respondnot only in color, and the front desk top made of rope cross compiled "arbor" reachmaterial on the echo, balance space indifference, virtual reality in.

Office Space 办公空间

"互融"：博炜曼办公空间
"Interaction"：BWM Office

项目地点：广东省广州市天河区中汇保利广场22F / Location：22F, Zhong Hui Bao Li Plaza, Tianhe District, Guangzhou, Guangdong
项目面积：320平方米 / Area：320 m²
公司名称：菲灵设计 / Organization Name：Feeling Design
设 计 师：伍文 / Designer：Evan Wu

IAI 设计优胜奖

菲灵设计 Feeling Design
中国 China

菲灵坚信每个空间都具有专属"灵魂"！
正如人们无法直接定义灵魂那样，菲灵无法直接描述"空间"的实质概念。
作为空间设计师，他们具有更加敏锐的触觉，通过单一或者多样的设计手法，将无形的"空间"转换为特定的"场所"。
菲灵专注于商业空间、展览空间及品牌终端环境的设计——设计源于感觉，一切从心！

Feeling Design believes that each space has its exclusive soul! They cannot describe the actual concept of space, just as people cannot identify soul. As spatial designers, they transfer "intangible space" into "functional space" with more acute sense and through one or more design approaches. Feeling Design devotes to giving each space an exclusive soul, offering people a totally new sensory experience and creating collision and resonance between people and the space. Feeling Design focuses on the design of commercial space, exhibition space and brand environment-inspired by feeling, follow heart.

人与人关系越趋冷漠的今天，在接到博炜曼的设计任务开始菲灵就力求突破。不受拘束，从属于自身的尺度，每个空间都有自己特殊的弧度，简洁空间与环保人造石形成的流畅线条交织出独特的形态。空间中白色与木质的契合，其柔韧性伴随隐藏式的柔和照明，给办公空间创造更多的融洽与和谐。

设计开始，设计师本能的追求一种简约但又具有惊喜的空间体验。流畅线条与简洁空间交织形成的美感，空间中白色与木质的契合，伴着隐匿式灯带映出的柔和照度给办公环境带来了舒适、安逸的氛围

Human relations are more hasten today, Feeling Design attempt to break through this situation after receiving office design task of Boweiman.Uninhibited, in this office, each space has its own special radian, the smooth lines of compact space and artificial stone interweave to a unique form. The combination of white color and wooden material, and the flexibility of hidden soft lighting, creating more harmony into the office space.

From the beginning of the design process, designers instinctively seek a simple but surprising space experience. The combination of white color and wooden material, and the flexibility of hidden soft lighting, bring easy and comfortable into the office space.

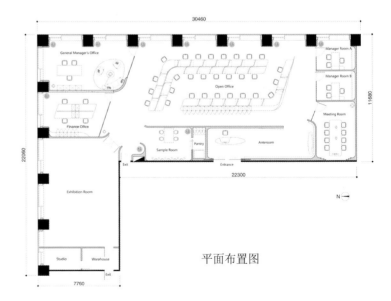

平面布置图

不受拘束，从属于自身的尺度，像这些空间的部分，每个都有自己特殊的弧度，以独特的形态相互交接和贯穿。中央开放办公区中相互流动的弧形办工桌挡板进一步加强人与人之间的互动性。

Uninhibited, in this office, each space has its own special radian, connecting and running through it in a unique form, and the curve insulation of office tables, enhancing interactive of people.

Office Space 办公空间

在对博炜曼的设计中,菲灵以弧形架构做基础,试图让空间里的所有元素相互融合,人造石、木质点缀,辅以柔和灯带,当然,还有融在舒适环境中的人。最终呈现出让办公者与来访者不免赞叹一声的作品,这小小的惊喜感,也是博炜曼带来的感受。

In Boweiman, Feeling design taking curve structure as foundation, trying to combine every element together, such as artificial stone, wooden material, and soft lighting, and the most important part, people. Finally finished design projects which bring surprise to employees and visitors, and this feeling, that's the feeling of Boweiman to desinngers.

在对博炜曼的设计中,菲灵以弧形架构做基础,试图让空间里的所有元素相互融合,人造石、木质点缀,辅以柔和灯带,当然,还有融在舒适环境中的人。最终呈现出让办公者与来访者不免赞叹一声的作品,这小小的惊喜感,也是博炜曼带来的感受。

三银集团上海总部办公室
SanYin Group Shanghai Office

项目地点：上海市青浦区 /Location : Qingpu District, Shanghai
项目面积：3 000 平方米 /Area : 3 000 m²
公司名称：上海亚邑室内设计有限公司 /Organization Name : YAYI (shanghai) Co.,Ltd.
设 计 师：孙建亚 /Designer : Alex Sun

IAI 设计优胜奖

孙建亚 Alex Sun
中国台湾 Taiwan, China

2009年至今　上海亚邑室内设计有限公司主持设计师；
1988 年　台北复兴美术工艺科毕业。

2009   Design Director of YAYI (shanghai)Co.,Ltd;
1988   Graduate of Fu-Hsin Trade & Arts School.

本案是功能性极强的办公空间，空间中设计了中轴对称式的长廊布局。结合建筑结构，在室内做了挑空和天井设计，针对自然光在不同时段和季节的光照对空间产生不同的环境效果，沿着中轴线分别在走廊两端各设计了圆廊和圆形天井。同时在中轴线上，沿着走廊还设计了一条长30米的挑空区，顺着水平方向和垂直的相互贯穿，增强了互动性和趣味性，也让空间更加多元化，使传统办公室的严肃、直白的氛围，在这里完全被颠覆。

This is a case of highly functional office space,and space design of the axis symmetric corridor layout. Combined with the building structure, in the indoor do pick empty and patio design, according to the natural light in different time and season of light produced different effects on the environment of space, along the axis respectively in at either end of the corridor design circle and circular courtyard Gallery, also in the central axis, along the corridor is designed a 30 meter long pick empty area, along the horizontal and vertical penetration, enhance the interactive and fun, also let a space more diversified, make a serious, straightforward traditional office atmosphere, here is completely subvert.

清润、淡雅的色彩，简洁的线条，把浩渺连绵的现代东方意境表现得恰到好处，令每一处细节都体现出清幽的氛围，从而营造出了与众不同的办公体验。

本案业主是中国制漆行业的佼佼者，把展示不同漆面质感的空间放到了二楼，设计师结合二楼透视感相当强的一面长廊墙面作为企业产品展示墙，既满足了产品展示效果，同时也增添了这个空间更为震撼的气势。

Qingrun elegant colors, simple lines, the modern oriental artistic conception vast sweep of the right, make every detail reflects the quiet atmosphere, thus creating a unique experience of office.

This case the owners as leader of a Chinese paint industry, to show different paint texture space on the two floor, the designer with two floor clairvoyant feeling quite strong side corridor wall as the enterprise product display wall, not only to meet the product display effect, but also add to this space more momentum shocking.

Office Space 办公空间

匡威研发中心
CONVERSE R&D CENTER

项目地点：广东省中山市 /Location : Zhongshan, Guangdong
项目面积：1 950 平方米 /Area : 1 950 m²
公司名称：京玺国际股份有限公司 /Organization Name : Exclaim United Corp.
设 计 师：周谦如 /Designer : Joy Chou

IAI 设计优胜奖

周谦如 Joy Chou
中国台湾 Taiwan, China

2014 年 陶朱隐园艺术住宅招待会所规划设计（台湾 台北）；
2013 年 健鼎科技品牌整合设计及展示厅规划（湖北 仙桃）；
2011 年 LESS 杭洲思创服饰江南布衣集团（浙江 杭州）；
2010 年 Esquivel 好莱坞时尚明星品牌（北京 瑜舍酒店）；
2009 年 五叶神 烟草品牌年度广告设计及品牌整合（广东 深圳）；
2009 年 大百汇集团盐田企业总部设计（广东 深圳）；
2009 年 大百汇集团品牌整合设计（广东 深圳）；
2009 年 统一银座 Unimart 超市品牌整合设计（山东 济南）；
2008 年 Converse China R&D Center 中国 Converse 设计研发中心设计（广东 中山）。

2014 TAO ZHU YIN YUAN Reception Office Planning Design, Taibei,Taiwan;
2013 Tripod Technology Corporation Brand Design and Planning Design of Exhibition Hall,Hubei,Xiantao;
2011 LESS,Hangzhou,Zhejiang;
2010 Esquivel,Beijing;
2009 Wu Ye Shen Advertising Design and Planning Design, Shenzhen,Guangdong;
2009 DABAIHUI GROUP Headquaters Design,Shenzhen, Guangdong;
2009 DABAIHUI GROUP Brand Design,Shenzhen,Guangdong;
2009 Unimart Brand Design,Jinan,Shandong;
2008 Converse China R&D Center,Zhongshan,Guangdong.

本案为知名鞋业集团设立于广东省中山市的 3 层楼的研发基地，一楼为办公区及样品间；二楼为研发中心生产线；三楼为企业总部。京玺国际团队将一楼的功能规划为：办公区、会议室、样品间、材料室、展示厅、主管室、接待区、会客室、VIP 休息区、休闲咖啡区等区域。

考究品牌本身悠久的历史文化，设计团队植入"殿堂""博物馆"的设计概念，展现品牌的潮流特色与时尚前卫印象。把在全球拥有广大忠实消费者的品牌，以前卫、个性的角度，诠释出其特有的潮牌文化。除了研发人员办公区域外，以宛若进入朝圣殿堂般的导览方式规划，包含休闲咖啡区、VIP 休息区、展示样品区，表现品牌的悠久文化与前卫特色。

CONVERSE R&D CENTER, Charisma Palace that sweeps the world.The case is a prestigious shoe group founded in the R&D base of 3 floors in Zhongshan City, Guangdong Province. The first floor of the building is an office and sample room, the second floor is the CONVERSE R&D CENTER and the production line, while the third floor is the corporate headquarters. Exclaim United Corp designs the area of first floor into: office, meeting room, sample room, material room, showroom, supervisor office, reception area, reception room, VIP room, and cafe. Appealing in the historical culture of the brand itself, the design team embeds the design concept of "palace" and "museum" by exhibiting the styling characteristics and fashionable avant-garde impression of the brand. Despite of its vast number of loyal customers worldwide, the brand is interpreted through particular stylish culture based on the avant-garde and personal perspectives. Apart from the R&D office, the team presents the room through guidance to sanctuary palace that not only includes a casual café, VIP room and showroom but also features the prolonged culture and advanced characteristics of the brand.

以品牌视觉形象做为区域主题，不同于传统的封闭型区域。以丰富鲜艳的颜色搭配间接光源，白顶棚延伸至立面，借由光影的变化演绎前卫时尚的品牌价值与概念。运用鲜艳颜色与光影的融合，借由视觉感官的刺激，进入品牌文化中传递潮牌活力精神。借由这样的连续、延伸，构建出场所的主题精神。

Plying the visual image of the brand as theme of the zone, the team incorporates indirect lighting with rich and bright colors to extend from the ceiling to the facade.

Different from the traditional sealed zone, the changes in lighting and shadow interpret the brand value and concepts in avant-garde fashion. Such fusion of bright colors and lighting penetrates into the brand culture to convey the spirit of style vigor though the stimulation of visual senses. The consistency and extension have collectively constructed the theme spirit of the field.

Office Space 办公空间

鼎天装饰办公空间
Ding Tian Decoration Company

项目地点：福建省福州市晋安区东浦路 133 号 /Location : No.133, Dongpu Road, Jin'an District, Fuzhou, Fujian
项目面积：1 050 平方米 /Area : 1 050 m²
公司名称：福建鼎天装饰工程有限公司 /Organization Name : Fujian Ding Tian Decoration Co., Ltd.
设 计 师：黄婷婷 /Designer : Huang Tingting

IAI 设计优胜奖

黄婷婷 Huang Tingting
中国 China

高级室内设计师；
福建鼎天装饰工程有限公司创始人、总设计师；
擅长项目：教育机构、办公空间、展示空间；
设计理念：设计是给予项目最适合的展示环境及空间舒适的体验。

Senior Interior Architects；
Chief Designer and Founder Fujian Ding Tian Decoration Co., Ltd.；
Great Projects:education institutions,office space,exhibition space；
Design Concept:Design is to provide the most suitable display environment and comfortable space experience for people.

本案作为一个装饰设计企业创意聚集的办公空间，设计师充分结合了自然元素和人文办公的需求点，淡雅的色调使整体空间更为的祥和安静，让人身心格外的放松。
在空间处理上运用纯粹的线条对整体空间进行分割，大面积的白橡面板、整面玻璃、木栅格的应用，色调统一，和谐的过渡让空间若即若离，零散、开放但又不失整体性。
设计师在入口和通道都采用大面积的云朵拉灰大理石材作为装饰，略显粗犷的表面和水泥墙形成有机整体，在灯光的映照下，使空间显得深邃而富有神秘的韵味。洽谈区应用整面的玻璃让空间融入周围环境，在保证私密性的同时又不显局促。为了延续空间的统一风格，主体办公区在布局上开放且井然有序，公共区域的座椅以白、绿为主色调，醒目而又不失协调。

As a work space where inspiration comes up, the designer makes full use of the natural elements and cultural office demand and puts it into a perfect connection. A peace and quiet atmosphere with quietly elegant tonal can make a person physically and mentally relax all the more.

Using pure on dimensional processing line to divide the whole space, bedding face of white oak panels, the whole grid application of glass, wood, tonal unity, harmonious transition space is compromise, scattered open but it does not lose its real.

At the entrance and channel, the designer introduces marble stone for decoration by large area. Slightly rough surface and concrete form an organic whole. In the backdrop of the light, a deep and rich space appears with mysterious charm. The application of the whole glass make the meeting room into an atmosphere with ensure privacy while not cramped. In order to continue a unified style of the space, the public areas of the seat is given priority to with white, green color, bold and do not break coordination.

Office Space 办公空间

Samlee 办公室
Samlee Office

项目地点：广东省广州市天河区 /Location : Tianhe District, Guangzhou, Guangdong
项目面积：480 平方米 /Area : 480 m²
公司名称：锐意（广州）设计 & 共和都市（香港）设计 /Organization Name : Ray Evolution Design Ltd&Republican Urban Design Ltd.
设 计 师：黄永才 /Designer : Ray Wong

IAI 设计优胜奖

黄永才 Ray Wong
中国 China

锐意（广州）设计有限公司创办人、首席设计师
共和都市（香港）国际设计有限公司创办人、首席执行官。

Chief Designer and Founder of Ray Evolution Design Ltd.;
Chief Executive Officer and Founder Republican Urban Design Ltd.。

Samlee办公室主张去繁从简的东方美学工作方式，符合了高速发展的城市消费模式，在信息高度运转的当下社会，本案诠释了城市群体与工作互动和个体间的关系：动与静，透明叠加，渗透留白的暧昧关系

本案是从城市新陈代谢的极简主义引申的结果，其核心的"动线"如蜿蜒的河流，人"流淌"于动线正契合"起、承、转、合"的东方美学。曲折蜿蜒的动线贯穿整个功能与形式，其余部分"留白"，折线与三角形的透明叠加，渗透模糊了空间界限关系，人与流动的时间在四维上的对话形成了二元辩证关系。

在空间布局上其核心是贯穿原建筑三个单位的动线，把公共空间，敞开办公室区域到半封闭的办公空间有机组合一起，打破原有的呆板空间布局。由动至静，在人随动线与立面折线的叠加移动过程中，形成了空间的暧昧性、故事性。

Without fussy details, the SAMLEE OFFICE was designed by a simplicity oriental aesthetics. This concept matches with the speedy developing city. In this highly running information society, the project presents the interactive relationship between the city, work and people – a kind of intimately relation of activity and inertia; transparent overlay; permeation blank.

This project is a result of extending minimalism from the city metabolism. The 'core line' just like a meandering river, the people flow in it, corresponds to the oriental aesthetics of 'opening, developing, changing and concluding'. The twist centre line through up all the function area. A special 'blank-leaving', with the triangle transparent fold line, overlaying and infiltrating the space, build up a four-dimensional binary relations between people and time.

The main theory of this layout is running through the three original architect units. The concept breaks the original formalistic space, but assembles the public area, opened-up office, and semi-enclosed office organically. From dynamic to quiescence, the users move in a streamlined plane and fold lined elevation, presents the ambiguous and communicated interior space.

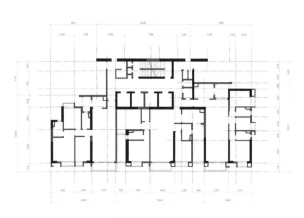

原始建筑资料图

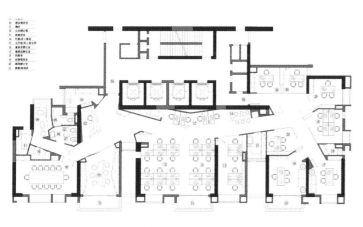

家具平面布置图

初·原色办公空间
Creative Origin

项目地点：台湾省桃园市三民路二段 257 号 /Location : No.257, Sanmin Road, Taoyuan, Taiwan
项目面积：170 平方米 /Area : 170 m²
公司名称：金湛室内设计有限公司 /Organization Name : Goldesign Studio
设 计 师：凌志谟 /Designer : Ling Chih-Mo

IAI 设计优胜奖

凌志谟 Ling Chih-Mo
中国台湾 Taiwan, China

2007 年创立金湛空间设计研究室 Goldesign Studio，致力于透过业主的内涵与现代风格反映出生活的空间时尚品位和文化层次，作品崇尚简约，线条比例充满美感。

透过创意延伸居住者的生活精神与内心期望，跳脱既定符号的框架与设限，逐一引导居住者从情绪酝酿至心境内化，透过界限之间的转折层次，触动个人深藏的记忆想念，寻求生活中种种美好的可能。

Ling Chih-mo established Jin Zhan Interior Design Co., Ltd. In 2007. He wants to show the taste of life space and culture according to the owner's characteristic and modern style. The feature of his work is simple, graceful lines.

Through the creative, extend the residents' spirit of the life and desire, jump out of limits and restrictions, guide habitant from brewing mood to internalization. By way of cross of different limits, touch people's deep memories, visit all sorts of beautiful in the life .

初·原色是是整体空间第一个想法，希望创意能一直停留在"饥饿的状态"。空间的分割规划充满灵活性，一楼前方的设计工作区是一个可以独立亦可以群聚的工作场所，可以滑动的办公桌结合之后就成为小组讨论的大会议桌。

墙面设计是希望带给空间人文意象，活动书架构成的长条形的阅读长廊，让知识、灵感可以跟着动线前进。另外透空的大理石楼梯，减少了隔绝感。楼下的绘图区也能和楼上有所互动，形成自由讨论区、灵活的阅读空间、浪漫的工作区。

"Creative Origin"is the first idea coming from the whole space ,hoping creativity can "stay hungry".The layout of the space shows flexibility. In the front of the afirst floor, the designer's workspace is suitable for both individual and team work. For example, the sliding office tables can be combined to become a big conference table for group meeting.

The wall was designed for bringing about humanistic images tothe space. The reading gallery composed of sliding bookshelves makes knowledge and inspirations move forward with the route. In addition,open-riser marble stairs reduce the sence of alienation and allow drawing area downstairs to interact with free discussion area,flexible reading space,and romantic workspace upstairs. to interact with free discussion area, flexible reading space, and romantic workspace upstairs. It is the main point of this design.

Office Space 办公空间

木制光影
Sunlight in the Wood

项目地点：浙江省宁波市鄞州区天童北路 1501 号麒麟大厦北楼 /Location : 0907-0908, Qilin Building, 1501 Tiantong Bei Road, Jinzhou District, Ningbo, Zhejiang
项目面积：200 平方米 /Area : 200 m²
公司名称：浙江宁波九筑空间设计有限公司 /Organization Name : Ningbo Jiuzhu Space Design Co., Ltd.
设 计 师：陈传畅 /Designer : Chen Chuanchang (Ateven Chen)

IAI 设计优胜奖

陈传畅　Ateven Chen
中国　China

宁波瑞时整体家居设计师；
宁波九鼎装饰有限公司主任设计师；
宁波阡陌装饰设计师设计总监；
宁波煊研装饰设计合伙人；
宁波九筑空间设计有限公司设计总监；
具有室内设计 5 年工作经验。

Designer of Racechina Company ;
Director of designers of JDZS Company;
Designer director of Qianmo Decoration Company;
Partner of Xuanyan Decoration Company ;
Designer director of Ningbo Jiuzhu Space Design Co., Ltd.;
Work as Interior Designer in 5 years .

有这么一个传说：钻木取火，它需要可挑剔性的材料、专一的动作和一颗恒心。借用这个传说传承易学文明故取名九筑空间。本案让设计师尝了回甲方的滋味。不局限于所谓的风格与手法，不局限于外界的主观因素，发出自身内心的需求，体现在选材、用材、颜色、个人偏向、喜好等方面。满足于自身的前提下，必须使得该空间：内部环境与外部环境的相互影响与和谐存在，促使人与环境的相互影响与和谐存在根本在于设计以人为本。

Focused on "Use original material, show transparent and bright feeling", this design applies symbolic expression, and multi design elements. The whole space combines Feng Shui and design in perfect way and implies company will have a unique and competitive development. Walking into office, firstly you can see the entrance with "loft" culture theme, surrounding white painted bearing channel steel and red brick, which create a strong visual shock. On floor, you can see the cobble and bluestone board. It expresses special idea and bright feeling. To the sitting room, design in casual theme and form a comfortable feeling under cold and warm harmony. The spatial space shows wide depth vision. With warm and bright color rendering, with the splendid concept, this style sets the company off a unique design. Next to the designer office area, the style is based on log wood, which gives you an imagination. It can take designers back to the origin and inspirit them new idea. The whole office has a balance between soft and hard material. Finally, the embellished lights echo the design theme, to show you a brand spirit combining unique design and practicability.

Office Space 办公空间

简谧·上海锐驰总部
Simplicity · Headquater of Shanghai Camerich

项目地点：上海市青浦区赵巷嘉松中路5369号吉盛伟邦国际家具村 /Location : No.5369, Zhaoxiang jiasong Road, Qingpu District, Shanghai
项目面积：545平方米 /Area : 545 m²
公司名称：十方圆国际设计工程公司 /Organization Name : S.F.Y. Design Engineering Corporation
设 计 师：赖建安 /Designer : Botta Lai

IAI 设计优胜奖

赖建安 Botta Lai
中国台湾 Taiwan, China

2011年 成立秀茶事连锁茶文化体系国际公司。
2007年 成立集慧堂软装设计工程公司；
2002年 成立十方国际设计工程公司；
1998年 成立安之居建筑师事务所；

2011 Establish An Zhiju Architecture Firm;
2007 Establish Ji Hui Tang Decorations display Design Engineering Company;
2002 Establish S.F.Y. Design Engineering Corporation;
1998 Establish Xiu Chashi chain tea culture system international company.

简单的柏拉图式形体空间，一个向心性的空间量体，以极简的方式诠释。以新鲜、纯粹、简单、健康的白色为空间主色调，增添了形式上的新颖、光影的变化。外立面采用大面积的留白艺术，框三五窗景，求其空灵，虚实相映，达到"无画处皆成妙境"的艺术境界，如此简单却给人以巨大震撼，让人回味无穷。

This is a kind of centripetal space organized by geometric solids in a simple, platonic style, conveying rich and complicated meaning in a very simple manner. Its main color is white, which embodies freshness, purity, simplicity and health, and in the meantime, it displays various expressions, bringing a sense of novelty to the whole structure and witnessing the change of light. The exterior facade, with several windows in it only, applies the art of leaving blank space in large area, to achieve a sense of tranquility and incredibility and to interweave the tangible and the intangible, reaching the artistic state that "blank space yields wonderland." Therefore, in spite of its simplicity, it makes shocking and memorable impression on people.

人字构造的楼梯支点，非刻意的矫揉造作，是以人为出发点，关注实用性，借由几何图形的穿插、变形、还原，重塑空间连接关系，加强现代量体概念。横竖条窗借景，让空间错位，透过光线洗练，使景致融合，虚虚实实、相得益彰。依梯而上，进入展场，依然延续人文概念，实现空间划分的灵活性与适应性，结合展示功能，引导动线，横纵空间依序展开，近、中、远景相互演变，引各自光影，呈多元感受与交流。

The pivot points of stairs are designed in an inverted "V-shape", which is not an act of affectation. Instead, based on the philosophy of people-orientation and the practicality, it is to reshape the connection of the objects in the space and strengthen the concept of modern space by virtue of the interweaving, transformation and restoration of geometric solids. The view-borrowing through windows of horizontal and vertical bars, dislocations of space, simplicity of the penetrating light, and the merge of varied views, all the tangible and the intangible come together, complementing each other. Go up the stairs, and here comes the exhibition hall, which continues to express the people-oriented philosophy, realizing the flexibility and adaptability of the spatial division. Combining the function of exhibition and directional movement line, the entire hall spreads orderly in length and breadth. And the medium close-up, medium-shot, and long-shot displayed evolve gradually, producing mixed feelings in people and diversified communications among them.

麦道置业办公空间
The Office Space Of MAIDAO Realestate

项目地点：浙江省杭州市余杭临平南苑街103号麦道大厦七楼 /Location : 7F Building Maidao , No.103, Nanyaun Road, Linping District, Yuhang, Hangzhou, Zhejiang
项目面积：2 500平方米 /Area : 2 500 m²
公司名称：中国美术学院 /Organization Name : China Academy of Art
设 计 师：王海波 /Designer : Wang Haibo

IAI 设计优胜奖

王海波 Wang Haibo
中国 China

2010年至今 浙江亚厦股份副总设计师、中国美术学院国艺城市设计研究院 副院长第九研究院院长；
2004—2009年 浙江中和建筑设计有限公司，设计总监中国建筑装饰协会高级室内建筑师、高级景观设计师、浙江省创意设计协会理事长；
2003年至今 中国美术学院讲师；
1999—2003年 中国美术学院环境艺术专业本科。

Since 2010 Deputy Chief Designer in the mansion co., LTD., zhejiang province,vice President of the Urban Design and Research institute,the China Academy of Fine Arts skill,the Dean of the Ninth academy;
2004—2009 Design director in ZheJiangZhonghe Architecture Design Ltd,Senior architect in Chinese Building Decorate Association,Senior Landscape Designer,President of design committee in Zhe Jiang province;
2003 until now Lecturer in China Academy of Art;
1999—2003 Bachelor degree, major in environment art design in China Academy of Art.

朴实的材质、简洁的线条、几何的造型、沧桑的老陈设与时尚的西方家具在此空间融合。虚实相间的隔断墙体隐现出多重的办公空间，直棱木栅与青砖墙透露着儒商的那份闲适与文雅，地面不同的材质界定出了办公区域、交通空间和休闲等候场所。包容、互通、内敛、简约是该办公空间特有的气质。主要材料采用仿旧大理石、橡木、竹板、青砖、玻璃。

The Office Space Design Description of MAIDAO Realestate's Unadorned material, concise line, geometric design and old school furnishings mix together with fashion western furniture in this space. Straight -edged palings and gray brick brick walls reveal the leisure, quiet and elegance of a scholar-businessman. Different textures on the floor define office area, circulation space and room for leisure and waiting,which is full of inclusiveness,intercommunication, connotational and uncomplicated soul in the office space. Main material:faded marble,oak, black brick,glass.

Office Space 办公空间

康华室内设计办公室
Samson Wong Design Group Office

项目地点：香港福源广场 /Location : Fuyuan Square, Hong Kong
项目面积：122 平方米 /Area : 122 m²
公司名称：康华室内设计 /Organization Name : Samsonwong Design Group Ltd.
设 计 师：黄业端 /Designer : Samson Wong

IAI 设计优胜奖

淳朴胡桃木与水泥、曲宛玲珑疏条隔屏、利落黑框线，与康华坚信的粗体之美互证。骄人4米楼高划分成无穷空间层次，炫酷船形会议室若空中楼阁。工作区开放、富有律动感，给予客户专业信心。

One can tell how the Office actualized its motto of "The Beauty of Bold" simply by stepping into the interior composed with the pureness of hickory and rawness of the cement,which synchronized with the absolute of the black wavy decors at the entrance. Professionalism of the group also revealed in the use of space in the entrance .Profession of the group also revealed in the use of space in the four-meters-high office, contained within it a cruise cabin ike meeting room where creative plans are born, giving out dynamics and openness where workers and visitors find accommodating.

康华室内设计
SAMSONWONG DESIGN GROUP LTD.

康华室内设计于2006年成立第一间室内设计公司，凭借时尚独特的设计风格和专业细心的服务态度，公司在短短数年内的发展已具相当规模。公司拥有专业且充满热诚的设计团队和项目顾问工程师，设计过的住宅项目很多。康华室内设计的设计顾问服务也包括国际知名餐厅、酒店和上市公司物业。

Samson was established in 2006 the first interior design company, with unique design style and fashion professional attentive service attitude, the company has developed quiet a scale in a few short years.
The company has a professional and energetic design team and project consulting engineer.It has service for innumerable projects.The design consultancy service including international famous restaurant,hotel and the list company's property.

黄业端 Samson Wong
中国香港 Hong Kong, China

曾就读于澳洲新南威尔斯大学修读室内建筑设计专业和澳洲新南威尔斯大学修读视觉设计专业。
Samson 于 2006 年创立 SAMSONWONG DESIGN GROUP LTD.。
Samson 一直贯彻多元化设计风格，勇于创新，忠于核心概念。为自己的专业宗旨。并因此深受各界赞赏和支持，历年来口碑不绝。

Samson studied in new south wales in Australia to study Interior architecture design,and in the university of Sydney in Australia to study visual art;
He founded SAMSONWONG DESIGN GROUP LTD.in 2006.
Samson has been carrying out diversified design style, innovative, loyal to the core concept for professional purposes,and was supported and appreciated by all circles, therefore, receive great review from allover the world.

清水混凝土考验工匠手艺，同样是 Samson 钟爱的材质，于工作间墙身大面积铺设，营造出真率不造作的柔和层次感，亦可解读为对建筑大师安藤忠雄的含蓄致敬。抬头仰顾，三盏金咖色丹麦枝竹金属吊灯明亮照人，为胡桃木独有的朴素沉实平添瑰丽，巨型灯饰恰足够骄人楼高比例，另有深咖色橡木墙身、顶棚及围屏作衬托；蜿蜒波浪形阶梯、假顶棚旁边设有职员工作柜，促进沟通流畅，工作氛围亦随之变得轻松跳脱，弯曲长梯所用的每片胡桃木，都经细意拣选。

The craftsmanship on the fair—faced concrete massively applied on the walls is a witty touch that adds perspectives sense of practicality and layers to the Office, such material that gained favours from Samson the Director,is being used in the design as a tribute to Master Tadao Ando ,the architect who expresses wonders of subtlety.Three golden brown metal chandeliers with the shape of the Artichoke shine like a Trio of staged musical, revealing the beauty of the hickory constructs as well as giving out the perfect balance between the deep brown oak walls, the ceilings and the screens. The streamline composed by the wavy stairs, the decorative ceilings an the desks of the workstations propels the vivid plans and proposals produced by the team. In the lively working environment with a staircase made with each piece of hickory seriously selected.

Office Space 办公空间

VIPABC
VIPABC

项目地点：上海市浦东陆家嘴银城中路太平金融大厦 5F /Location : 5F, Peace Finance Building, Yincheng M Road, Lujiazui, Pudong District, Shanghai
项目面积：2 800 平方米 /Area : 2,800 m²
公司名称：大衡建筑设计 /Organization Name : DH Architects
设 计 师：陈威宪 /Designer : Weisen Chen

IAI 设计优胜奖

陈威宪 Weisen Chen
中国台湾 Taiwan, China

大衡设计有限公司 建筑设计总监；
1997 年 任职美国 GLC 设计事务所；
1992 年 任职大砚建筑事务所；
1993 年 成立陈威宪建筑师事务所。

Architectural Design Director in DH Architects;
In 1997, work in U.S. GLC design firm;
In 1992, work in Dayan architecture firm;
In 1993, founded Chen Weixian architect firm .

因为互联网，人类对于信息的获得，从来没有这么开放而自由，但也因为互联网，人类也走到另外一种封闭的世界。设计师觉得互联网，正在积极改善着世界，也正在无情地摧毁这世界。因此设计师想利用空间特性找回办公环境的人性化尺度，平衡工作与生活的心态，创造一个时尚、温馨并有创造力的工作环境。入口大堂，代表互联网的无限延伸。利用六面镜面反射的空间，延伸无限的 LED 格子，让入口的视觉冲击达到某种晕眩的程度，虚拟的超大空间，这就是互联网的世界，没有尽头，也看不清尽头，但交织穿越的却是立即可达的画面和光线，没有确切的地点和方位感，一切是如此的丰富，却如此地孤独，这样虚幻的世界是设计师对互联网的无限想象。

Thanks to the Internet, the acquisition of information has never been so open and free; however, it leads people to another isolated world in the same time. The Internet is destroying the traditional interpersonal relationship while improving the world in a positive way; therefore, designers idea is to restore the balance of human nature, the balance between work and life, by creating a warm, modern and creative working environment.
Entrance Lobby, the infinite extension of internet. In order to create a world resembling the world of internet, designers use mirrors on all six planes in the space to reflect out endless LED grid, which blurs the boundary of the room, and brings visual impact on people stepping into it. People lose their senses of direction and position once seeing nothing but grid of light and scenes on the displays in the lobby; it's a maze out of designers' imagination of Internet.

Office Space 办公空间

因为现在的搜索技术存在太多商业目的，如果使用者习惯用固定的 IP 在固定的搜索引擎上使用，网络会逐渐记录我们的习惯和喜好，改变使用者搜索的排序，因此使用者得到的答案会越来越贴近自己的思路，反而让资讯获得越来越狭隘而主观，所以改变这种行为，最好的办法就是不要有固定的电脑，而且最好不要在固定座位上工作。设计师企图让上班变得有趣，所以空间呈现不同尺度的桌椅和形态，大家可以每天因为团队分组作工作位置的调整，并且到处分布讨论区、沙发区和写字板，家具的颜色、材质都显得非常的自然温馨，且色彩丰富。

这是追求高科技的团队，但在人文上的诉求也同样重要，所以在会议室、公共大堂的现代科技表现上，设计师增加了许多创意，例如指纹叶状的树、大幅海报和色彩鲜艳的讨论区等，就是想超越冰冷科技的印象，表达人们对生活的热情，在生活态度上更有艺术的气质。

The technology of searching engine involves too many commercial purposes nowadays, if users are used to use fixed IP to search on fixed searching engine, their behavior and preference will be noted and recorded, and then gradually the searching results will become narrower and more subjective, because the searching engine will alternate the results according to their preference in the past. The best way to stop searching engine from doing so is not offering fixed computers and fixed seats to employees.

The owner of VIPABC accepts this concept immediately because the culture of the company is to develop most advanced network technology while maintaining the logic of human thinking.Designnners attempt to make ordinary daily work more interesting, so they create a dynamic working space full of discussion corners, sofas, and whiteboards, where employees can team up and work in different place with furniture of different scale and style every day.

Office Space 办公空间

Villa Space
别墅、豪宅空间

South Lake Mountain Villa	158
OCT No.10 Villa, PuJiang	162
Hongmei 21	164
M House	166
Zen Industry	168
The Spring of Milan Yang Fu	170
A Pledge to Live with Sea & Mountain	172
Extend，Extending	174
Photosynthetic Breathing Curtilage	176

158	南湖山庄别墅
162	华侨城10号院D户型样板房
164	虹梅21
166	M住宅
168	工业味的禅意
170	米兰春天杨府
172	海誓山盟
174	延续、延续
176	光合呼吸宅

南湖山庄别墅
South Lake Mountain Villa

项目地点：广东省广州市南湖山庄 /Location : Nanhu Village, Guangzhou, Guangdong
项目面积：355 平方米 /Area : 355 m²
公司名称：Danny Cheng 室内设计有限公司 /Organization Name : Danny Cheng Interiors Limited
设 计 师：郑炳坤先生 /Designer : Danny Cheng

郑炳坤 Danny Cheng
中国香港 Hong Kong, China

毕业于加拿大多伦多国际营销与设计专业；
2002 年开设 Danny Cheng 室内设计有限公司；
郑氏设计的设计一直贯彻简约、建筑美感和空间规划为设计重点；
郑氏设计范围主要围绕住宅、样板房、精装修设计和重点参与香港大部分发展商项目。近年亦开始接触中、港、澳房地产的新开发项目，如公共空间、会所和大型销售厅。

Having graduated from International Academy of Merchandising and Design in Toronto, Canada;
He established his own company -Danny Cheng Interiors Limited in 2002;
Danny's designing notion has always been one of simplicity, with strong architectural aesthetic and precise spatial layout;
Except for serving private clients, his work also covers residential flats, show flats and the design of decorative rooms.In recent years, he starts involving in modern real estate development schemes in China, Hong Kong as well as Macau, designing a variety of show flats, public areas, club houses and large-scale sales office.

这是一座 4 层高的别墅，位于著名的南湖风景度假区旁，设计师把方案打造成悠闲写意的度假屋，以融合于周边翠绿茂密的大自然环境。不仅在高度上拥有得天独厚的优势，再加上设计师赋予其时尚特别的内外造型，使这两种元素在设计中交织融化。在远处观望，这座山顶大宅给人远离城市喧嚣的距离感。设计师以简洁的线条、独特的木条子旋转门、木条子屏风、错落有致的阳台和空中悬挑泳池等几何空间设计，赋予半山建筑独一无二的态度和内涵。当迈进大门的一刹那，又会重新审视这座别墅，会发现在它高傲的外表下掩藏的是一种富有亲和力的时尚感，因其内部很高，看起来充满建筑感。

This four-storey villa is situated beside the renowned International Private Resort. In order to perfectly integrate the holiday villa into the greenish natural environment surrounding it, the designer ingeniously created a flat which brings forth a cozy and relaxing feeling. Apart from the unique advantage of its extraordinary height, Danny Cheng also granted the villa a modern and astonishing internal and external appearance. These two design elements integrated into one flawlessly. In the distance, the holiday villa on the peak looks like somewhere far away from the hustle and bustle of the city centre. The designer made use of some geometry elements in this exceptional design, like some simple lines, a distinctive revolving door with wooden stripes, a partition with wooden stripes, a balcony with irregular patterns and an infinity swimming pool, granting the villa on the hill some individual attitude and aptitude. One will be impressed by a completely different feeling the moment he steps through the main door, while he discovers a kind of modernized affinity hidden behind its unapproachable appearance. The building looks architectural because of its particularly great internal height.

室内设计遵循自然、简约主义，设计的元素、色彩、照明、原材料都被设计师简化到最少的程度，但这种设计风格对色彩、材料的质感要求很高，所以简约的空间设计非常含蓄。入口处，设计师以灰色和黑色的大理石地面点缀墙面，配以木质原色的地板，使得整个建筑设计充满了自然的味道。

The internal design follows the principles of being natural and simple. All the design elements, uses of color, illumination and raw materials are simplified by the designer to the largest extent. However, this kind of designing style requires a very high standard on the choice of colors and textures of materials. As a result, the simple-styled spatial design is extraordinarily subtle. At the entrance, the designer especially decorated the wall with some grey and black marble and matched this with a wooden floor of original color, filling the entire design with a kind of natural sentiment.

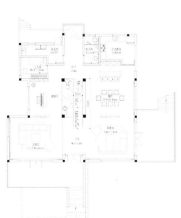

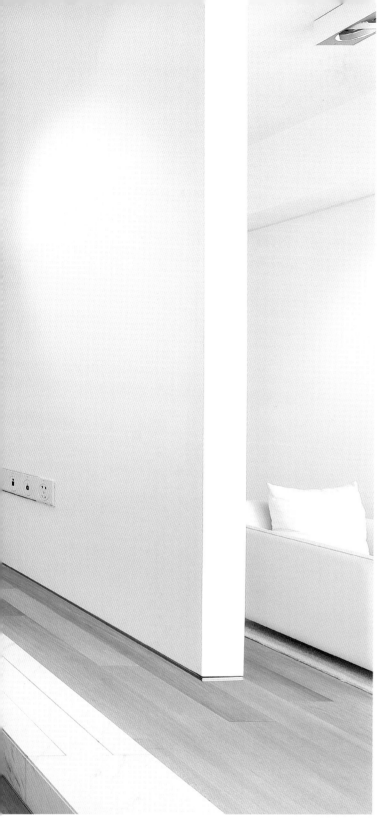

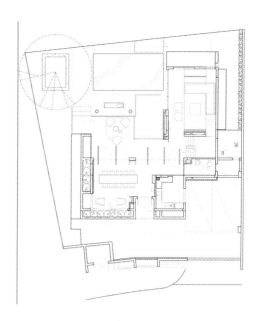

同时这两种不同材质的搭配也点亮了整个空间,当阳光从窗户投射进来,光线会从大理石的墙面反射到木质的地板上,形成流动的线条,滋养着旁边的绿色植物,使得整个空间充满了自然的气息,同时也能让业主自然的进行角色的转换。

At the same time, the perfect match of these two distinctive materials also lights up the whole area. While sunlight is pouring in through the windows, the light will be reflected from the marble wall onto the wooden floor, creating some beautifully flowing lines which are nicely connected with the nearby green plants, making the entire space like part of the nature. In this way, the owners of the house could easily and naturally imagine the change of their roles.

华侨城 10 号院 D 户型样板房
OCT No.10 Villa, PuJiang

项目地点：上海市闵行区 / Location : Minhang District, Shanghai
项目面积：650 平方米 / Area : 650 m²
公司名称：上海亚邑室内设计有限公司 / Organization Name : YAYI (Shanghai) Co.,Ltd.
设 计 师：孙建亚 / Designer : Alex Sun

IAI 设计优胜奖

孙建亚 Alex Sun
中国台湾 Taiwan, China

2009年至今　上海亚邑室内设计有限公司主持设计师；
1988 年　台北复兴美术工艺科毕业。

2009　Design Director of YAYI (shanghai)Co.,Ltd;
1988　Graduate of Fu-Hsin Trade & Arts School.

"光"与"影"，看似虚无，却可以赋予建筑和空间真正的魅力。本案是一栋富有文雅气息的独立别墅，展现出一派奢华禅意的风格。日光，毫无遮掩地穿过窗框和玻璃照了进来，华丽感在客厅的石墙上跳跃，停在客厅金属与粗犷石材相互融合的装饰墙上，这就是高挑的共享空间，开阔的厅堂所带来的视觉盛宴。

"Light" and "shadow", seems empty, but can give the building and space real charm. This case is a rich and elegant atmosphere of the independent villas, showing a pack of luxury Zen style. Sunlight, there is no cover through the window frame and glass shining through the gorgeous feeling, from the living room walls, stop the mutual fusion in the living room and a rough stone wall metal decoration, shared this is tall, open hall brings the visual feast.

虹梅 21
Hongmei 21

项目地点：上海市闵行区 /Location : Minhang District, Shanghai
项目面积：650 平方米 /Area : 650 m²
公司名称：上海亚邑室内设计有限公司 /Organization Name : YAYI (Shanghai) Co.,Ltd.
设 计 师：孙建亚 /Designer : Alex Sun

IAI 设计优胜奖

这是一个老别墅改造项目，整体设计包含了建筑外立面改建部分。老房子本身存在的空间结构和建筑外立面具有不合理性。本案为坡屋顶别墅，改造设计成极简的建筑风格。设计师对建筑及外立面进行了较大修改，把原有的斜屋顶拉平，并且把外凸的屋檐改建为结构感很强的外挑，以方盒为基础的设计理念，重新分割成性能较强的露台或雨篷，既增强了建筑的设计感，又增大了空间的实用性。从户外景观、建筑，一直到室内，一气呵成，没有间断和多余的装饰。外墙窗户成为设计过程中非常重要的一环，所以尽可能地扩大窗户的范围，并且避免出现一切多余的框线，把所有外墙窗框预埋隐藏在建筑框架内，达到室内外没有界限的效果。

This is an old villa reconstruction project, the overall design of building facade renovation part contains. The old house itself unreasonable spatial structure and building facades. This case the owners background for overseas fashion advertising creative people, the owners advocating minimalist. A twenty-year old villa roof slope roof, to transform the design into a minimalist style of architecture. The designer of buildings and facades were larger changes, the original inclined roof leveling, and rebuilds the convex roof structure with a strong sense of the cantilever, design philosophy to box based, resegmentation success can strong terrace or awning, can enhance the sense of design construction, and increase the utility of space. From the outdoor landscape, architecture, has been to the interior, minimalist spirit must be coherent, without interruption and superfluous decoration. The windows in the external wall becomes very important in the design process, so as far as possible to expand the scope of the window, and avoid all the extra frame lines, all exterior window embedded hidden in the building frame, no limits to indoor and outdoor.

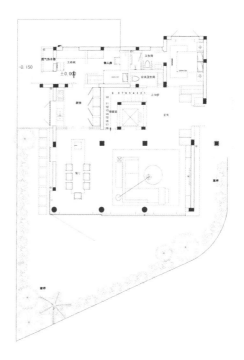

M 住宅
M House

项目地点：香港西贡区 /Location : Xigong District, Hong Kong
项目面积：237.2 平方米 /Area : 237.2 m²
公司名称：安信室内设计工程有限公司 /Organization Name : Anson Cheng Interior Design Ltd.
设 计 师：郑秋基 /Designer : Anson Cheng

IAI 设计优胜奖

郑秋基　Anson Cheng
中国香港　Hong Kong, China

郑秋基作为安信室内设计工程有限公司设计总监，从事室内设计和工程管理业务超过 15 年，参与超过 200 项设计项目，参与多项独立屋及复式单位设计工程，获得多项国际设计大奖，Anson 及其团队一直以" 以客为本、团结及创新 "为宗旨，每次设计，均会先了解客户的喜好、要求、生活习惯和品位，凭借丰富的设计经验和潮流触角，通过颜色、灯效、材料的配搭，以及空间上的策划和工程时间的控制，务求为客户打造心目中理想的空间及设计。
公司提供多方面室内设计和顾问服务，范围包括住宅、办公室、店铺、展览会、陈列室及品牌发展等。

As design director of Anson Cheng Interior Design, Anson engages in interior design and project management business more than 15 years, and participates in more than 200 projects.He takes part in a number of design projects involved house and loft, and wins a lot of international design awards. "Client first, unity and innovation" are the purposes of Anson and his team. At the beginning of design, they comprehend requirements, living habits and tastes of their clients. With rich experience in design and tide antennae, through color, lighting effects, material supplement, time and space planning and project control, they create an ideal space and design for customers.
Company provides various services of interior design and consultancy, including residences, offices, shops, exhibition, showroom and brand development, etc.

M 住宅是位于香港西贡的独立屋，此屋用作度假及与朋友聚会之用。
设计环保简约，以白色作为主色调，配合少许活泼的颜色点缀，与大自然亲近而加入天然木元素，再结合时尚的灯光艺术，充分利用独立屋的大面给与高楼的优势，将整个空间变得简洁而大气。合理地安排一些独特的装饰品，让每一个角度都展现出不平凡的一面。设计上，以突出单位可吸纳日间自然光及景致的优点，让每区均有开敞的空间，亦同时节约能源。客厅采用白色及鲜黄色的元素，为简约设计增加色彩，另加入可变换的 LED 灯，营造不同气氛，以增加趣味。开放式厨房上的仿天窗设计，加上餐厅的木制家具，令空间充满自然气息。

M House is a fun holiday home that has been consciously designed to be ecofriendly. Responding to their client's brief to save on energy the building shows how mezzanine spaces and primary colors – in this case – white, can be used to draw sunlight in through its skylights and windows, meaning every last drop of daylight is captured to stave off switching on. Designed with a few to the winter as well as summer months the team has added touches of bright, lively colors and other little artistic embellishments such as LED lamps here and there to add a strong sense of personality and warmth that ensures this space is anything but bland.

工业味的禅意
Zen Industry

项目地点：澳门高士德大马路 3C-3DC/Location : 3C-3DC, Dama Road, Gaoshide District, Macao
项目面积：2 955 平方米 /Area : 2 955 m²
公司名称：昊设计工程有限公司 /Organization Name : ET Design &Built Ltd.
设 计 师：谭沛嵘 /Designer : Eric Tam

IAI 设计优胜奖

谭沛嵘　Eric Tam
中国　China

"昊设计工程有限公司"简称 ET Design。
"ET Design"是于 2003 年成立的公司，云集港澳设计师，是专注于高端住宅、商业、办公环境策划的装饰公司，10 余年成为一家拥有一流管理、高水平设计、精湛的施工和完善的售后服务一体化的专业化品牌装饰公司。

ET Design &Built Ltd. is also named ET Design, ET Design was founded in 2003, at the same year ET Design & Built Ltd. was established. ET Design gathers a lot of designers from Hong Kong and Macao.It's a decorate company which focus on the high-end residential, commercial, office environment planning .After 10 years, ET Design becomes a professional brand decoration company which has a first-class management, high level design, excellent construction and perfect after-sales service.

业主是一位工程师，他喜爱着工业味道的东西，但也想令家变得温暖及带点禅意。在本案，设计师主要的目的是抛开传统的风格，混合简约现代风，令整件事变得简单又可以带出主题，抛开传统的中式花纹，利用材料上的颜色配搭出最贴切的禅意感，与业主的意见互相结合。
当大家走进屋内，第一眼就会见到通往阁楼的黑铁制作成的楼梯和鱼池，也是全屋最"抢眼"的地方，然后可以慢慢静下来坐在楼下的鱼池边脱下鞋子，专注地望一望鱼池内的小龟及小锦鲤，更是我们最想营造出的设计效果，也令业主每晚回到家时的一个心灵过渡，给自己时间去欣赏生活氛围中的美事。

The owner is an engineer, so he loves things with industrial flavor. However, he also wanted a warm home with zen sense. In this case, designers main purpose is to abandon traditional styles and to mix simple contemporary styles and highlight a certain subject. In consideration of the opinion of the owner, they dismiss traditional Chinese decorative pattern, and resort to the matching of material colors to obtain the most appropriate zen sense. Stepping into the house, you will catch the first sight of fishpond and black iron-made stairs leading to the attic, the most eye-catching place. You may first feel surprised, and then calm down slowly, sit beside the fishpond, take off shoes and stare at the turtle and fish inside the fishpond. This is the very effect they strive to achieve in the design. It will bring the owner atmosphere to relax once he returns home, making him slow down his pace to enjoy his life more joyfully.

Villa Space 别墅、豪宅空间

米兰春天杨府
The Spring of Milan Yang Fu

项目地点：福建省厦门市米兰春天 /Location : The spring of Milan, Xiamen, Fujian
项目面积：1 200 平方米 /Area : 1 200 m²
公司名称：厦门东方装修设计工程有限公司 /Organization Name : East Design Decoration Engineering Co., Ltd.
设 计 师：曾冠伟 /Designer : Zeng Guanwei

IAI 设计优胜奖

曾冠伟　Zeng Guanwei
中国　China

注册高级室内建筑师；
厦门东方设计装修工程有限公司设计总监；
厦门东方设计研究院副院长；
厦门室内装饰协会设计委员会主任；
U 空间室内设计师国际联盟副主席；
美克美家管理学院 培训师；
厦门地产奥斯卡凤凰花奖专家评委；
厦门别墅家居联盟别墅设计名家；
意大利米兰理工大学国际室内设计管理硕士；
清华大学酒店设计高级研修班结业。

Registered Senior Interior Architect;
Xiamen East Design Decoratin Eng.Co. Ltd.,Designer Director;
Xiamen East Design Institute Vice-dean;
Xiamen Interior Decoration Association,Director;
U-Space Interior Designer International Alliance Vice-chairman;
Makorhome College of Management Trainer;
Xiamen Real Estate Oscar Phoenix Flower Award Expert judges;
Xiamen House and Home of Villa Alliance Villa design master;
Politecnico di Milano University Master of International Interior Design Management Hotel design of senior seminar of Tsinghua University Completion .

本案地处风光秀丽的白鹭洲筼筜湖畔，背山临水，景色宜人，是一栋得天独厚的豪宅。但是空间在功能上存在不足，楼梯入门太近，没有电梯不利老人生活。设计师进行了大刀阔斧的改造，最大限度地将自然景观融入室内。业主是一个三十多岁的年轻精英，少年老成不喜现代风格，独好中式新古典及地域文化并尊重设计。因此，设计师确定本案风格为"混搭新古典"。

This case is located in the beautiful scenery of the egret Yundang lake, mountain water, scenery and pleasant, is a set of be richly endowed by nature in the penthouse. But there are a few problems,the stairs from door is too close and no adverse old elevator life etc..The designer of the reform will make snap.Let the maximum natural landscape into the interior.The owner is a thirty-year old young elite,an old head on young shoulders do not like modern wind,only good new classical Chinese style and the region culture and respect design. In view of this, designers determine the case mix of "neo classical style".

家具也是中西合璧，有业主收藏的中式家具，有美克美家、卡萨贝拉的欧式家具，有鸿风堂的家具，还有设计师针对特殊空间设计的订制家具。空间里有业主收藏的古董、艺术品、陈设品；虽是中西混杂，但在设计师的精心平衡之下，一派和谐，彰显大气富贵。四层还设计了一个中式的木构佛堂，与整体空间也一脉相承，相映成趣。

Furniture is "Chinese and Western", the owners of Chinese style furniture, "Markor", "Casa Bella" European style furniture, "Hong Feng Tang" colonial style furniture, and designers for special space design custom furniture; the owners of antiques, works of art, the Chen Shepin space; is also the Huayang hybrid, but under the carefully balanced design division, a harmonious, rich atmosphere. The fourth layer has also designed a Chinese style wooden temple, and the whole space is come down in one continuous line, gain by contrast.

海誓山盟
A Pledge to Live with Sea & Mountain

项目地点：台湾淡水 /Location : Danshui, Taiwan
项目面积：602 平方米 /Area : 602 m²
公司名称：鼎睿设计有限公司 /Organization Name : DINGRUI Design Studio
设 计 师：戴鼎睿 /Designer : Tai Ding-rui

IAI 设计优胜奖

鼎睿设计有限公司
DINGRUI Design Studio

公司一路走来十几个年头，设计用心、认真地看待每个案子，希望每个案子都可以表达出真、纯、净、沉、极的初心。简单、自然、自在、唯一、无所顾虑的心，作品注重人与空间之情感．

原名戴啃峰空间设计有限公司；
1998 年成立迄今，于 2008 年 1 月更名；
设计总监，戴鼎睿（Ray）。

The company over ten years, designers treat every case so seriously ;
They hope every case can show the willing of true,pure,calm,extreme.They want to show heart of the simple ,nature ,easy,only. The works pay more attention on the emotion between people and space;

Old name is Dai Dingfeng space design company ,change the name in 2008.1;
Design Director Ray.

人与空间、光影的对话，留白给人一个好的背景，无彩色加上线条与板块，重组这个空间的氛围。
项目外在坐拥难得的山海景观，内在则拥有对生活细腻的品位。被它深深吸引，原来空间是宁静的音符在那跳跃着，为了成就那份第一眼的平静与感动。设计师知道应该让它属于回归，利用大自然给予的环境，让空间可以无限延伸生活的气度，极简的灰与白让窗外的色彩涉入，舒服的用材可与窗外的自然连结，室内的可用空间与室外的可观景色，让业主在泡澡的同时也享受着宁静的美好

A dialog among human, space, light and shadow.Blank is a great background which has no colors with only lines and blocks to reconstruct the feeling of space.
The house is located on a hillside of Tamsui. The external surrounding has rare mountain and sea views and the internal space is required by an exquisite taste in living.The designer was deeply attracted since the day of seeing the place. He felt the space just like a quiet note jumping over there. In order to accomplish peace and move at first glance. He realizes he has to return it to the natural environment, using only gray and white to involve more colors from outside windows. And, the comfortable materials could be connected with outdoor nature and indoor space. You may enjoy the movement of quiet and peace when you are taking a bath with the useful space and beautiful landscape at the same time. So the space can be infinitely extended to your life.

延续、延续
Extend, Extending

项目地点：台湾省新北市板桥区庄敬路 199 号 16 楼 /Location : 16F, No.199, Zhuangjing Road, Banqiao District, Xinbei, Taiwan
项目面积：167.38 平方米 /Area : 167.38 m²
公司名称：本入设计有限公司 /Organization Name : Sidstory Design
设 计 师：黄仁辉 /Designer : Sen Huang

IAI 设计优胜奖

黄仁辉　Sen Huang
中国台湾　Taiwan, China

2012 年至今　本入设计有限公司；
2008—2012 年　齐观设计有限公司；
2005—2006 年　鸿新室内设计有限公司；
2002 年　文化大学建筑及都市设计学院；
2000—2002 年　成城建筑师事务所。

2012　Founder of Sidstory Design;
2008—2012　Qiguan Design Co.,Ltd.;
2005—2006　Hongxin Interior Design Co.,Ltd.;
2002　School of Architecture andUrban Design and Planning, Chinese Culture University;
2000—2002　Chengcheng Architects.

家是一代传一代延续故事的地方，隐喻"树"来延续家的生命力。当爱的种子种植在土地上、发芽茁壮后形成了树形，树形抽象延伸的线条进入室内象征了生命由外到内的延续，并生长成挑空楼板的线条编织成家的轮廓。

树形抽象延伸的线条勾勒出工作及居住平面的范围界线，直梯与折梯的相遇。不管在哪里都可以感受到家人的动向。设计师利用温暖材质的历史感铺陈了家的回忆，视觉的穿透力及连贯性丰富了立面表情。一条走廊产生亲子互动的力量，紧密联系了两个空间，分享故事和玩耍时光，无论何时，都能参与孩子们成长的每个瞬间。

Home is a place where generations and memories are continuously extended, one after another. The tree is a metaphor implying vital extension of the home. The seed of love is planted, sprouted and grown into trunk.The trunk abstractly grows into the home as it signifies life is extended from outside to inner space. The shape of the atrium flooring is formed, which becomes the contours of the home. Use of natural elements creates an aging-like home filled with memories. With addition of visual transparency and elements unity, this makes a heart warming home.The hallway not only strengthens parent-child interaction but also closely links these two separate spaces.Whether story time or playtime, parents will not miss a child's growing moments that are simply priceless.

Villa Space 别墅、豪宅空间

光合呼吸宅
Photosynthetic Breathing Curtilage

项目地点：台湾省桃园市 /Location : Taoyuan ,Taiwan
项目面积：486 平方米 /Area : 496 m²
公司名称：青埕空间整合设计 /Organization Name : Clearspace Architecture and Interior Design
设 计 师：郭侠邑 /Designer : Ryan Kuo

IAI 设计优胜奖

郭侠邑 Ryan Kuo
中国台湾　Taiwan, China

青埕空间整合设计有限公司负责人 主持设计师；
青埕空间整合设计建筑与人文空间研究；
设计特点为生态自然 * 当代艺术 * 人文时尚；
一直以来把绿色建筑概念，试着融入到各层面的空间设计当中。服务内容着重于建筑整合、空间设计、私宅订制、生态绿色建筑、节能住宅和养身会所，以及商业空间、医疗办公、餐饮酒店、家饰美学和视觉形象整合规划设计。

Director of Clearspace Architecture and Interior Design;
Clearspace Architecture and Research on Humanistic Space;
Design Features are ecological nature,contemporary arts and human fashions.ClearSpace Architecture and Interior Design with the concept of green building has tried to integrate in diverse space design. It focuses on the design of architectural integration,space design,private residence,green building,energy-efficient residence,health club,commercial space, medical office,restaurant and hotel,decoration aesthetics and visual brand design.

旧建筑、旧格局，原本是非常狭长的空间。从开阔、延伸的角度，垂直水平的轴线、虚实轻重的比例，在线性轴线下链接空间舒适而开放的尺度，6.1米长的客厅、串连厨房和餐厅，形成家的凝聚力中心。看着由天井洒落的阳光，听见潺潺流水声、感觉空气中的温湿度、触摸环保天然材质的质感、用心去感受这一切，源自于光合效应的五感生活。建筑体中段的天井设计，引入自然光，让整个狭长的空间明亮起来，并配合空气塔的概念，收到环保省电的效果。景观水池流水的设计，有效地降低室内的温度。室内种植半日照的绿色植栽能吸收二氧化碳排出新鲜的氧气，植物的光合作用，也达到空气净化的目的，创造自然有氧、身体心灵长居久安的五感生活。

This old building is a outdated layout and very narrow long space.From the open, extending angle, vertical and horizontal axis, the ratio of solid and empty, light and heavy, the linear links the comfortable and an open space. More than 6 meters island connects living room, kitchen and dining room. It is the cohesion of family.Looking at the sunlight floating down from the skylight, listening to the gurgling sound of water, feeling the temperature and humidity in the air, touching the texture of natural materlals, dlligently feeling and enjoying it all, bring all five senses into life.The skylight is in the middle of the building. It brings in natural light into this entire narrow long space. With the air tower concept, it meets environmental friendly energy efficiency.Water flow in landscape design can effectively reduce the indoor temperature. Indoor plants can absorb carbon dioxide and discharge fresh oxygen. Both plants and people carry out photosynthesis. Not only purify the air but also create a natural oxygen, body and soul be long life of five senses.

Apartment Space
公寓空间

Black DNA	180	黑色 DNA
Initialization	184	初始
Graphene	186	墨方
Great Mind	188	器宇
The Temperature of the old House	192	旧屋的温度
Guo's Residence	196	台中郭宅
Huang's Residence	198	台中黄宅
Chen's Residence	200	台中陈宅
Coexistence	202	共生心钛
Environment · Tranguil	204	境·静
Greenview Flow	206	绿波
Light into the Scenery	208	和光沐景
Yonghe Hsu House	210	永和徐宅
Ocean Heart 3	212	海洋都心 3
Forest City	214	中半山独立屋
Pure Wood	218	淳净木沐
Grassroots	220	草根清境
Lighting Technology · Fashion Aesthetics New Look	222	科技光感·时尚美学新面貌
Garden View of Yuanzhuang Manson	224	景上苑庄邸
Poetry Colors	226	赋采
Light Flow	228	光影流转
S Residence	230	石宅
Contemporary Art Residence	232	当代艺术宅
Rounded House	234	圆舞
Foshan Dong Yi Beauty in Design	236	佛山东怡美居设计

Apartment Space 公寓空间

黑色 DNA
Black DNA

项目地点：台湾省新北市 /Location : Xinbei, Taiwan
项目面积：368 平方米 /Area : 368 m²
公司名称：玮奕国际设计工程有限公司 /Organization Name : Wei Yi International Design Associates
设 计 师：方信原 /Designer : Fang Xinyuan

IAI 特别奖　IAI 年度最佳设计机构

玮奕国际设计工程有限公司
Wei Yi International Design Associates
中国台湾　Taiwan, China

转化·演绎　设计向度

玮奕国际设计工程设计总监方信原认为，设计团队就空间设计规划而言，已具备独到而且广度的创意特色。
在 2014 年初，跳脱传统以家具为主的装饰表现，要将设计的触角伸及艺术，以研究、非制式、量产为主，建立空间独创性与深度，遂成立 C&F 艺术研究工作室。

为了让设计更多元化、国际化，融入不同国家的特色，将事业版图扩及上海，成立 L&F 联合建筑设计事务所。无论是设计的触角或是事业的版图延伸、茁壮，都使玮奕国际设计团队以无比的热忱与态度，在专业的设计领域里，继续坚持！

TRANSFORMATION, TRANSLATION, DIRECTION
" When it comes to design scale, Wei -Yi Design has precision, creation and individuality. We have shifted our priority from furniture decoration to art. So that we established C&F Art Studio in early 2014, more focus on research and production." said Fang, Xinyuan, the creative director of Wei-Yi Design.

They are supposed to involve more point of view and international aspect into design, and that's why they expanded their business to Shanghai and established L&F Architecture Design Firm. And when they have talented people from all over the world, it might as well cause some impact to Taiwan in every way. No matter what they do and how they do, Wei-Yi Design is devoted to it with 100% of their heart.

本案位于台北大都会地区，因在设定和一般住宅有所不同。故案件空间结构的特殊性，使得本案空间设计规划上，产生了许多可能。
概念上寻找一个共同的语汇，这语汇就如同乐曲中美妙的音符，缭绕在整个空间里。折纸的艺术，巧妙地运用在这个结构中。而元素的运用，也在原生、多重、协调、层次等架构下，进行运用。犹如 DNA 的元素的不同排列，产生不一样的结果。

This project is in the Taipei metropolitan area. As its usage is different from most houses in general, these conditions, together with the uniqueness of its spatial structure, have generated many possibilities in the space planning.
Common vocabulary is sought for this concept, and this vocabulary is like the melodious notes in a musical composition, lingering in the entire space. The art of origami has been cleverly used in this structure, and the elements are being applied with the concept of native, diversity, coordination and layering in mind. Just like DNA, the sequencing of elements gives rise to different results.

通过极具特殊性、设计性的楼梯结构,将开放空间和私密空间贯穿起来。而建筑语汇的着痕,也无时不经意地在设计里流露出来。如天际线的设定、块状的置入、线条比例的分割等,均透露出运用建筑元素的构思来规划。室内置入于建筑中,而建筑置入于整个城市,彼此都是关连。而人是这些关连中最小的单元,也是最基本且最重要的元素。

The public and private areas are linked through the unique and carefully designed stairway. Evidence of the architectural vocabulary is inadvertently displayed in the project from time to time. For example, skyline, blocks and divisions through lines portray how the concepts of the architectural elements are used in the design. The interior sits within the building while the building is within the city, and they are all interrelated. Humans are the smallest element in this relationship, and also the most basic and important element.

Apartment Space 公寓空间

初始
Initialization

项目地点：台湾省台北市敦化南路 1 段 376 号 10 楼之 1/Location : 10F, No.376, Dunhua South Road, Taibei, Taiwan
项目面积：148 平方米 /Area : 148 m²
公司名称：玮奕国际设计工程有限公司 /Organization Name : Wei Yi International Design Associates
设 计 师：方信原 /Designer : Fang Xinyuan

IAI 年度最佳设计机构

本案位于台北近郊的"三峡"，由该名称，则可得知这个城镇的特殊地理环境。"山城"就成了它的另一别称。

This project is in "Sanxia", a town near Taipei. From its name, it can be seen that the town has a unique geographical environment, and "mountain city" has thus become its alias.

在空间规划上，明确区分为公共空间及私密空间两大区块，整体空间的调性，运用了材质的特性，将整个结构立面延伸贯穿，打破各功能区域的界限。运用单一色调的规划，多样化材质的变化，探讨材质之间的差异所产生的细微变化，如青砖和特殊水泥质感的差异，青砖和盘多摩之间的碰撞所产生的微妙变化。另一方面，回收木料、生铁等材质的运用，亦对环保概念做了进一步的探讨。而这些单一色调的材质，其微小变化上的运用，突显出本案规划上细致的一面，重新定义精致的涵意。空间大方向的规划，开放的手法，使得不同场域的机能及运用，独立但又融合。而单一色调，多样材质的使用，所表达的空间氛围寓意且诠释着另类的"禅"。

In terms of space planning, the public and private areas are distinctively separated. However, the tonality of the entire space makes use of the materials' characteristics to extend and connect the various structures, extending the definition of the various functional areas. The designer makes use of a monochromatic theme and the changes of the diverse materials to explore the differences between the materials and the minute changes they generate, such as the differences in the texture of blue bricks and special cement, and the subtle changes generated when the blue bricks collide with Pandomo. In addition, materials such as recycled wood and pig iron were used, adhering to the concept of environmental protection. Making use of the minute changes of these monochromatic materials highlights the delicate side of the planning of this project, redefining the meaning of exquisiteness. Also, the main direction of the space planning adopts an open concept, making the function and usage of the various regions independent but integrated. The ambience of space expressed through the monochromatic theme and usage of diverse materials has generated an alternative meaning and interpretation of "Zen".

Apartment Space 公寓空间

墨方
Graphene

项目地点：台湾省台北市 /Location : Taibei, Taiwan
项目面积：78平方米 /Area : 76 m²
公司名称：尚艺室内设计 /Organization Name : Shang Yih Interior Design Co.,Ltd.
设 计 师：俞佳宏 /Designer : Chia-Hung Yu

IAI 年度最佳设计机构

尚艺室内设计
Shang Yih Interior Design Co.,Ltd.

俞佳宏　Chia-Hung Yu
中国台湾　Taiwan, China

设计的艺术，取决于空间动线、收纳、实用方便性和风格做的完美的结合。
拥有16年设计及工程经历，具备室内设计乙级技术师资格。

设计精神
强调团队精神结合专业服务为主要目标，公司成员向心力强、坚持技术专业第一及服务优先两大诉求。

经营理念
注重沟通，尊重业主的需求，希望在空间功能与业主的需求间取得平衡。并将业主个人特质呈现其中。

The art of design depends on the perfect combination of space dynamics,collection and convenience and style.

Design Spirit
Emphasis on team spirit and professional services as main goal.

Business Ideas
Focusing on communication, respecting;
clients'demands,combining and balancing clients' demands and characteristics with space function.

本案的设计是为一位拥有诗书气息的单身女性细腻地描绘如书画笔墨一般，蕴含人文气息的居住空间。在格局配置上，强调简洁方正；在颜色氛围上，则着重于静谧与雅致；而整体的视觉观感，更是倾心于空间连贯的对称性。
以铁件的沉稳质感，将一块拥有大自然色泽与肌理的石材，围合成一个完美比例的长方框形；壁炉上的大理石，流水般的线条将公共空间的利落刻画出间接性的区隔；而拥有整个空间的核心价值。名为泼墨山水的大理石，将设置于客厅与书房、客厅与餐厅的中轴线上，沉稳地带出原木温暖的色泽，与拥有大地触感的空心砖墙面。

A lone woman is drawn in an ink painting style and portrays a feeling of poetry and literature. The living space focuses on being concise and angular; with regards to color and atmosphere, a quiet and elegant tone is portrayed, while the general visual perception stems more from the symmetry of the space.
The space hasa rectangular frame with perfect proportions and walls made of stone with a natural shine and smooth texture. Marble is placed on the furnace to create a slight visual partition of the orderly public space by using flowing lines. The color of the marble is called "splashed ink landscape" and its purpose in the overall space is to have its own warmth and luster play with the warmth of the wood on the axis between the living room and the study and the living room and the restaurant. The hollow brick face provides just a touch of earth.

Apartment Space 公寓空间

器宇
Great Mind

项目地点：台湾省高雄市 /Location：Kao-hsiung, Taiwan
项目面积：70 平方米 /Area : 70 m²
公司名称：尚艺室内设计 /Organization Name : Shang Yih Interior Design Co.,Ltd.
设 计 师：俞佳宏 /Designer : Chia-Hung Yu

IAI 年度最佳设计机构

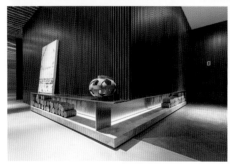

通过玄关木格栅区后，迎面而来的是不锈钢材质的轻食及餐厅区。大气好客的男主人热爱烹饪，设计上大胆地使用不锈钢材质于吧台区，呈现现代与科技感，流动的时尚光线，布满餐厅吧台空间。

冷冽的不锈钢材质，顶棚用温暖的木格栅来取得空间材料上的平衡，企图使轻食区域在五感中成为易亲近的厨房。

双十轴线动线配置，用开阔的空间迎人，小巧精致的私密空间置于双十轴线的后方，成为一家人的休憩之处，大气的居所，正落落大方地呈现这一家庭的气度。

After passing through the wood mullion area of the hallway, the refreshments and dining areas made of stainless steel are within sight. The generous and hospitable host has a passion for cuisine. The stainless steel is boldly applied to the design of bar to give the character of modernity and technology. The flowing light with fashion style is spread all over the bar space where the crowed is gathered.The cold stainless steel ceiling is leveraged by the warm wood mullion with the attempt of creating the refreshments area the most approachable kitchen for the five senses.

The wide-opened public area with the circulation of double cross axle is quite welcoming. The compact and exquisite private area where the family has a rest is situated behind the double cross axle. The extraordinary residence exactly reflects the grand bearings of the family.

Apartment Space 公寓空间

Apartment Space 公寓空间

旧屋的温度
The Temperature of the Old House

项目地点：台湾省台北市 /Location : Taibei, Taiwan
项目面积：72 平方米 /Area : 172 m²
公司名称：尤哒唯建筑师事务所 /Organization Name : TA-WEI,YU Architects
设 计 师：尤哒唯 /Designer : Ta-Wei Yu

IAI 最佳设计大奖

尤哒唯 Ta-Wei Yu
中国台湾 Taiwan, China

尤哒唯建筑师事务所主持建筑师；
聿和室内装修设计工程有限公司负责人；
大元联合建筑师事务所设计师；
东海大学建筑硕士；
建筑物室内装修专业技术人员登记证书。

设计理念
生活建筑 功能多样 人性绿能

Design Director of TA-WEI,YU Architects;
Director of Ya-Yu Interior Design Co.;
Designer of KRIS YaAO Artech Firm ;
Master of Department of Architecture of Tunghai University;
Certificate of Interior Decoration professional.

Design concept
Building life, Multifunction, Green humanity

此案是顶楼加盖的两层住室空间，采取"退缩"的手法，退让、加大了阳台、露台，将面向公园的光与景观引进、融入室内。利用回收的木材、工业生产的砖、水泥、铁件，以及保留铁皮屋桁架、旧墙的原貌，作为空间的铺陈。
"绿色"已融入生活，于是随处可见"可回收"设计如走廊上大小不一的格栅，重新演绎现代和室推拉门。回收桧木重新设计，用于儿童卧室的闻香主墙。

The transformation of a demoded flat with rusty rooftop addition to a chic, lukewarm maisonette, is conceived in the conception of "subtraction". By minimizing to the essentials, the house corners that are deluged with scenic views and natural lights in the living hearth yield to the augmented balcony and terrace for embracing the green horizons atop the metro park. Moreover, exploited with the industrial production of bricks, concrete and steels, the original walls and trusses are retained to structure the functionality of interior layout in elaborate textures.

Exceptionally, 'Green Living' permeates into the design details in terms of 'recycling'. First, the grilles in a variety of dimensions flank the corridor for a fresh interpretation of the contemporary sliding door.Next,the air in children's bedroom is full of the natural flagrance released from the recycled flakes of Taiwan cypress that are embellished on the wall.

顶楼的铁皮空间,则利用模块化、回收的旧栈板,填入三角形屋的既有空间里,装置成一堵兼具收纳、展示的起居主墙,装置栈板墙的剩料,做成了椅凳、茶几、床边柜及床架等家具设计;实木搭配七彩缤纷的墙体,也成为对室外大自然的记忆延伸与装置的连结。这是旧料再利用,并结合传统工法与保留旧屋记忆的空间展现。

Third, the rooftop space is triangularly shaped by metal sheets and partitioned with a wall decorate which are subtly structured by the recycled pallet modules. Interestingly, the interlaced gratings created by the modules extend the functionality in both exhibition and stowage. At last, the pallet remnants were further used in the making of furniture such as chair, stool, coffee table, bed stand and bed frame. Moreover, the variegated wall decorated with the reclaimed hardwoods even conveys the linkage to their lush flora in memory. Herein, the extensive use of recycled materials, the perfect integration of traditional techniques and the retained memory of dated house are accordingly embodied in every aspects of the interior settings.

194 Apartment Space 公寓空间

台中郭宅
Guo's Residence

项目地点：台湾省台北市西屯区 / Location : Situn District, Taichung, Taiwan
项目面积：40 平方米 / Area : 140 m²
公司名称：珥本设计 / Organization Name : Urbane Design
设 计 师：陈建佑 / Designer : Steven Chen

IAI 设计优胜奖

陈建佑 Steven Chen
中国台湾 Taiwan, China

陈建佑 + URBANE DESIGN
陈建佑，毕业于中原大学室内设计学系。
URBANE 是都会洗练、温文尔雅的形容词，这样的形容也十分符合我们作品表达出来的氛围，我们以珥本这两个字来表现思路脉络，珥为古代朱玉耳饰，玉乃自然优美之材料、耳则为人体构造极富细节之处，就像在设计的操作中，我们想呈现出材质最原始的特性，探讨其在空间中比例、分割、轻重、厚薄的关系，并经由空间分割排列的过程，达到机能与动线，光线与阴影，穿透与封闭的相对关系。

URBANE DESIGN
Graduate -Department of Interior Design ,Chung Yuan Christian University.
The "suave and urbane" atmosphere is what we exactly want to express. URBANE has two meanings. "Er" is a kind of ancient jade earrings, and the jade is graceful product of nature, while ear is one of subtle human body, as we would like to display the original characteristics of material in the design, discussing the relationship of proportion, division, weight and thickness, the relative relation between function and dynamics, light and shadow, penetration and closure in the process of space division and arrangement.

质朴、宁静、愉悦，回归家的本质是设计师尝试的规划主题。为使全家人生活起居，经由空间设计串连在一起，而进入私密空间后又可个别保有独立空间，所以在空间设计上注重连结性，用以形塑家的核心凝聚力，如将阅读及上网功能集中在公共区域，卧室则单纯以休憩为主。同时也须处理小面积住宅的局促和收纳问题，让使用者能在较大的空间尺度裡享受适意的行住坐卧。

The simple and unadorned,peace,and essence of returning home is the theme which design had try to plan.In order to care for the whole family needs,the living space is connection by the space design,and it can also keep the individual independent space once enter the private sector.So designers use the begin and end connection in space design to build the domestic core cohesion,such as concentrated the reading and internet function in the public areas.And the bedroom is simply functioned in rest. At the same time,it also deal the storing problem in small area,so the user can enjoy the comfortable living in bigger space.

Apartment Space 公寓空间

台中黄宅
Huang's Residence

项目地点：台湾省台中市国美馆周遭 /Location : Around National Taiwan Museum of Fine Arts, Taichung, Taiwan
项目面积：198 平方米 /Area : 198 m²
公司名称：珥本设计 /Organization Name : Urbane Design
设 计 师：陈建佑 /Designer : Steven Chen

IAI 设计优胜奖

针对退休后夫妻二人的使用对象，设计的空间尺度可以放大许多，并能更精准安排生活所需的功能范围。从玄关进来，以书房作为接待，客厅反为二人独享的隐私空间，并以浅色地砖和染灰木地板作为空间的转换分割。以在欧洲的旅居体验，选择了蓝绿色作为家具的主题，呈现出令人平和宁静的居家氛围。餐厅旁的开门为主卧入口，用同色调的蓝色为背墙配色，经过廊道式的更衣间分别为2间男女主人的卧室。因考量俩人年老后的使用习惯，采用无障碍的空间设计。

Targeting on users as retiring couples, spatial design scale can be enlarged lots, it also can make more precise arrangement of function range that is needed for life. After entering from the entrance, designers utilize study room for reception, the living room becomes private space for the couples, and light color floor tile and stained ash wood flooring is for converting space division. With experience in living in Europe, designers select blue and green as furniture theme presenting peace and quiet living atmosphere. The opening next to the restaurant is entrance of master bedroom, it has same tone blue as back wall color, what after corridor-style dressing room are two bedrooms for male and female masters. Because designers consider the living habit after couple is getting old, They appeal to accessible space.

台中陈宅
Chen's Residence

项目地点：台湾省台中市文心森林公园附近 /Location : Around Wen-Hsin Forest Park, Taichung, Taiwan
项目面积：175 平方米 /Area : 175 m²
公司名称：珥本设计 /Organization Name : Urbane Design
设 计 师：陈建佑 /Designer : Steven Chen

IAI 设计优胜奖

室内的格局走向，有别于大空间的连结处理，设计以破碎性的墙面组合，如客厅与餐厅交会的一道 L 形形清水模砖墙与功能柜结合，延伸同样语汇的方式至主卧室入口前的木墙端景，以错落为表现方式，避免空间上有被阻隔的感受。透过清水模砖材料的延展，更拉近立面之间的关系，使客厅与餐厅相互连结，并让光线自然流通。而在空间色调上，以大地与森林色系为主，并透过材料具备的纹理特质加以表现。

Indoor design pattern is different from connection treatment design for great space; designers utilize combination of shattered walls, such as connecting one L shape water mold brick wall with a functional cabinet at the cross of the living room and dining room, which extends using method of same vocabulary to the wooden wall side view in front of the main living room, that presents a scattered style to avoid making you feel being isolated in the space; through extension of water mold material, it can make relationship between standing surfaces closer, make living room and dining room connect each other, and allow light flow through naturally. In setting color of space, designers mainly apply earth and forest color, and utilize the texture characteristics possessed by materials to present the space.

Apartment Space 公寓空间

共生心钛
Coexistence

项目地点：台湾省台北市北投区行义路－天母富贵 /Location : Xingyi Road, Beitou District, Taibei, Taiwan
项目面积：70 平方米 /Area : 70 m²
公司名称：原境国际室内装修设计工程有限公司 /Organization Name : Yuanking Interior Design
设 计 师：邱郁雯 /Designer : Chiu Yuwen

IAI 设计优胜奖

邱郁雯 Chiu Yu Wen
中国台湾 Taiwan, China

建筑环境研究所硕士；
原境国际室内设计设计总监；
台湾室内装修工程管理乙级技术士；
高级室内建筑师。

Education Background Master of the Institute;
for Built Environment;
Design Director of Yuanking Interior Design;
Taiwan Interior Engineering Management Technician Grade B;
China Senior Interior Architect.

本案探讨着空间各式元素的沟通及包容性，意通过此案，向当代设计风格寻检介质、场所、人物的三角关系，看似迥然相异却实则能彼此通透并存的可能性。
介质：以自然触觉与金属温度，在"心"与"钛"的冲突字眼间，琢磨最前卫的材质关联。
场所：最大限度地连结空间的内外之景，借由光线形成出设计色彩与视觉美学无死角的表演舞台，同时作材质的载体，催生色光与材质交相辉映的光彩。
人物：以开放式的动线规划，演绎着尽管个体天差地别，却衷心追求着和谐的共生心态。同时借由介质、场所、人物的三方交叠，鼎立出稳固无虞的空间氛围。

The project probes into the connection and compatibility among various spatial elements and investigates the possibility of coexistence among the three vastly different elements-the medium, space and personage in contemporary design styles.
Medium: finding connections with avant-garde material by expressing the contrast between human beings and titanium with a natural touch and metal attributes.
Space: maximizing space by interconnecting interior and exterior scenes, and conveying the color design and visual aesthetics through lighting as the material reflects its intrinsic beauty and radiance in addition to creating a no dead angle live performing stage.
Personage: setting up an open hallway to present the pursuit of a harmonious coexistence in the living environment in spite of the differences among individuals. The interweaving of the medium, space and personage bolsters a Shangri-la like ambience.

Apartment Space 公寓空间

境·静
Environment · Tranguil

项目地点：台湾省台中市西区 /Location : West District, Taichung, Taiwan
项目面积：297 平方米 /Area : 297 m²
公司名称：大言室内装修有限公司 /Organization Name : Great Word Design Company
设 计 师：黄金旭 /Designer : Chin-Hsu Huang

IAI 设计优胜奖

黄金旭 Chin-Hsu Huang
中国台湾 Taiwan, China

大言设计 主持设计；
中原大学室内设计研究所 硕士；
中国科技大学室内设计系 助理教授。

Principal Designer of Great Word Design Company;
Master of Institute of Interior Design, Chung Yuan Christian University;
Assistant Professor of Interior Design Major, University of Science and Technology of China.

本案设计师与业主沟通之初，得知业主注重养生，因此设计师提出以"自然"为全案的发想主轴，在规划上运用天然石材纹理、原始木皮的颜色、流线造型顶棚的设计手法和户外造景等元素，使得内外空间相互呼应，展演出空间意境"山、林、水、景"，搭配微风天光的流动，演绎出雅致、静谧的空间氛围。

When the designer and proprietor of this project began communicating, the designer discovered that the proprietor is health-conscious. Therefore, the designer proposed "nature" as the theme of this project and utilized natural stone textures, the colors of raw veneer, a streamlined ceiling design, and outdoor landscaping in the design of this project. The internal and external environments reflect each other, representing the spatial concept of "mountains, forests, water, and landscapes"; combined with the flow of breeze and sky, an elegant and quiet spatial atmosphere is created.

在开放性空间架构里，主卧室入口墙面透过木质肌理，引至电视背墙及神明厅，运用纹理界定空间区域的划分，却能让它们无缝连接，赋予空间开阔的可能性。客厅大理石墙透过上层结构内缩完美地与木质界面作衔接，随着木皮延伸、转折至顶棚，利用木皮纹理使墙壁、顶棚相互交融，再延伸而下一直连接至餐柜区，构筑出三个功能各异的量体，使公共空间一气呵成。至于私密空间配置上皆运用纯粹的材质和色调，搭配简洁的细节来诠释本案空间意境。依据家庭成员喜好及功能，呼应家具式样，烘托场所整体氛围，营造出自然、环保的生活蓝图。

In the open-space architecture, the wood texture of the walls at the entrance to the master bedroom leads to the television panel wall and the ancestral shrine. Texture is used to define the boundaries of spatial areas, and also enables the seamless connection of each area, providing possibilities for open spaces. The marble walls of the living room converge with the wooden interface perfectly through the superstructure. The veneer extends and transitions into the ceiling. The veneer texture is used to merge the walls and the ceiling, and extends further downward to connect to the kitchen cabinet area, simultaneously creating three structures with distinct functions and forming a common space. Pure materials and hues are combined with concise details to derive the spatial conception of private space in this project. Furniture designs reflect the preferences and needs of family members to enhance the overall atmosphere of .

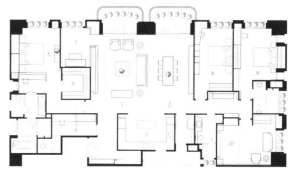

Apartment Space 公寓空间

绿波
Greenview Flow

项目地点：台湾省新北市新店区中央路 155 号 A 栋 22F-1/Location : 22F-1, A Building, NO.155 Zhongyang Road, Xindian District, Xinbei, Taiwan
项目面积：198 平方米 /Area : 198 m²
公司名称：格纶设计工程有限公司 /Organization Name : Guru Interior Design co., LTD.
设 计 师：虞国纶 /Designer : Frankie Yu

IAI 设计优胜奖

虞国纶　Frankie Yu
中国台湾　Taiwan, China

2004 年成立格纶设计有限公司，自创立以来室内设计项目遍及住宅空间、商业空间办公场所、会所样版房等，对于风格与品味的追求，强调人与空间的关系，以生活品位与空间艺术的角度，成就设计的独特性。
自幼学习美术，求学时期参与台湾绘画比赛并获奖无数，1990 年顺利完成学业进入设计事务所学习，负责项目别墅、住宅、大型知名饭店和办公场所、百货卖场等设计项目，历时十余年之久。

2004 Establish Guru Interior Design co., LTD. From the foundation, the projects of interior design consist of living space, commercial space, office space, club space etc. About the style and taste, Guru Interior Design Consultant focus on relationship between people and space. Form uniqueness of design from taste of life and art of space.
1990 Finish study, and work in several design studios. During about ten years, take charge of diverse cases such as villa, house, famous hotels, office space, shopping mall etc.
Study painting from childhood. At school, participate in a large amount of painting contests at Taiwan and receive many awards.

共生——阳光、空气、水，透过直接或抽象的形意转换并设计居住行为体验，不断由内外环境的构成系统中去思索或觉悟，从建造、定居、存在，反应对天地秩序的虔敬与感念，生活自然而然被转化为一种有机共生的仪式。
形态——和其光、同其尘。破除界线、融化整合、涵构引申、回归自然。
拟真——挹注绿意，回应阳光与空气的和谐律动，追求一种融合、共生、平衡的动态真性秩序。
意译——依水岸之傍，于粼粼波光间，享天光、拥绿意。
将设计的语言转绎为不拘泥的生活对话，在流动的空间映象里，顶棚层次蔓延并转化演译成曲度的水纹肌理，抽象的解译自然而然的隐敛，在介质、在表情、在虚实、在光影中，静默不语。

Coexistence——Through a direct or abstract meta-transition, elements of sunshine, air, and water were turned into the design of an architecture masterpiece which reflects the respect and appreciation of the order of the nature. Living in the architecture, a result of repeated thinking of the factors making up of the internal and external environmental systems is a coexistence with the nature.
Form and Style——The design pays equal emphasis to lights, form, and style.
Dynamic True Order——The design is to pursue a dynamic and true order consisting of harmony, coexistence, and balance with the nature by merging the green environment, sunshine, and air as design elements. It aims at the making of a living which is conducive to the inner peace of mind.
Understated Beauty——The design of the architecture takes the advantage of its proximity to the water, and its location to view the ripples, sky lights and the lush vegetation. The architecture is an interpretation of abstract and great design concepts into a dialogue with life materialized with the ripples of water, the dancing of the lights and shadows that speak of the understated beauty.

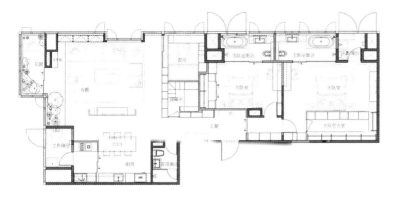

和光沐景
Light into the Scenery

项目地点：台湾省台北市阳明山格致路 203 号 5F-6 / Location: 5F-6, 203 Yangmingshan Gezhi Road, Taibei, Taiwan
项目面积：93 平方米 / Area: 93 m²
公司名称：格纶设计工程有限公司 / Organization Name: Guru Interior Design co., LTD.
设 计 师：虞国纶 / Designer: Frankie Yu

IAI 设计优胜奖

Apartment Space 公寓空间

永和徐宅
Yonghe Hsu House

项目地点：台湾省新北市 /Location : Xinbei, Taiwan
项目面积：65平方米 /Area : 65 m²
公司名称：上海大诠室内设计工程有限公司 /Organization Name : Miemasu Interior Design
设 计 师：卢国辉 /Designer : Lu Guohui

IAI 设计优胜奖

卢国辉 Lu Guohui
中国台湾 Taiwan, China

2007 年　成立大观室内设计；
2004 年　宏盛帝宝实品屋执行设计；
2004 年　上海金茂法式餐厅执行设计；
2002 年　任职动象国际室内设计。

2007　Establish MIEMASU Company ;
2004　Executive Designer of Baker House at Hongsheng Dibao;
2004　Executive Designer of Shanghai Jinmao French restaurant ;
2002　Work at Trendy Interior Design Co; Ltd.

家——该是一个能让家人无拘无束互动的空间，这是业主对于家的空间需求。因此公共区域以开放式的空间，让家人互动关系更为紧密，并利用简洁线条的高低顶棚设计，以无形的领域划分，连结公共空间的属性。

自然返家，就是使心灵最大解放，纯洁的白色，搭配温润色调木皮，赋予空间自然的调性，舒缓了城市中的纷扰。

沉静——阅读应是可使空间、时间静止的事。运用深色木皮，让整个空间弥漫着沉静的氛围，可专注地埋首于案牍之间，亦可或坐或躺于卧榻，沉浸在阅读的世界之中。

利落——生活该是简单自然，但不失秩序。利落的线条造型分割，以此为设计语汇，藏浴室入口于其中。

The owner of this residence request a comfortable space which can connect family interaction. In such opening space, the harmonious sense of serenity with livelihood can then be unveiled in great joy. Simple lines of ceiling creative high and low two vision design which defines the space into invisible.

Natural——Pure of white can relax mind when you we're come home. With many warm colorful veneer, the room projects a warm embracing quality and a relax, enjoyable and natural atmosphere.

Calm——Reading allows the space and time still. Using the dark veneer, that makes the entire space diffused quiet atmosphere, while I focused and immersed in the world of reading no matter sitting or lying on the couch.

Neat——Life is simple, natural but yet orderly,the style of neat is hiding in the entrance.

Apartment Space 公寓空间

海洋都心 3
Ocean Heart 3

项目地点：台湾省新北市 /Location : Xinbei, Taiwan
项目面积：132 平方米 /Area : 132 m²
公司名称：上海大诠室内设计工程有限公司 /Organization Name : Miemasu Interior Design
设 计 师：卢国辉 /Designer : Lu Guohui

IAI 设计优胜奖

本案位于新北市淡水区的临海样品房，业主期望打造此案为海滨度假居所，设计概念以此建案名称发想，将海洋相关的意象融入立面设计和材质选用之中。

This constructing project is a sample house on the waterfront , located at Dain-Sue district, New Taipei City. The constructor hope to make it become a seashore resort , so the design concepts come from the name of the project and get the sea-related imaginations (scenery) involved in the structure designed and the material chosen .

空间规划上，以休闲的接待会所为考量，将全室大半划分为公共区域使用，将客厅、餐厅、书房相互串连，自入口进来，即可感受到其大气的尺度，使亲朋好友于此交流互动更为自在。

On space arrangement, recreation reception office is mainly considered. Large half of the house will be used as public area with living room, dinning room and study room connected. From the entrance, the feelings of wide (spectacular or generous) space is obvious and let the relatives and friends doing activities here feel freedom.

立面设计以圆弧线板等距分割的语汇，传达出船坞夹板规律的线条，借此修饰厨房和儿童房入口暗门，搭配典雅壁纸，以重复的分割与材质，强化空间延伸放大的感受。壁面与顶棚选用桧木染色木皮，以仿旧刷色的纹理，表达海砂质感的意象。主卧床头主墙特别挑选仿马毛质感的进口壁纸，配合床头上方的吊灯，赋予起居空间犹如饭店般的精致质感。

The facial design is separated into same distance pattern with arc line boards to express a shipyard boards regular line feature, that could decorate the hidden entrances of the kitchen and the kid's room. Use the elegant wallpaper, cut repeatedly, and the material to strengthen the feelings of space extension and enlargement. The walls and ceilings choose hinoki-colored bark and mimic ancient paint pattern to express sea sand texture. The main bedroom head wall is decorated with specially selected imported wallpaper like horsehair texture. Having the chandelier above the bed gives the bedroom space like a fine quality of a high class hotel.

Apartment Space 公寓空间

中半山独立屋
Forest City

项目地点：香港中半山 / Location : Zhongbanshan, Hongkong
项目面积：133.3 平方米 / Area : 133.3 m²
公司名称：Tade 设计有限公司 / Organization Name : Tade Design Group Ltd.
设 计 师：黄世杰 / Designer : Kenji Wong

IAI 设计优胜奖

黄世杰 Kenji Wong
中国香港 Hong Kong, China

Tade 设计有限公司总设计师及行政总监。
Kenji 的设计理念在于重视空间感、生活品位、颜色和材料的配合与运用。
其作品简约而富灵魂，不建议业主采用过多装潢。
以人性化为主轴，依据业主个人风依环境现况条件，提供设计建议。

General Designer and the CEO of Tade Design Group Ltd.
The design concept of Kenji is that attaches great importance of applying to the space, life taste, color and material.
Their works are simple and spiritual. They don't suggest that clients use too much decorates.
They provide design suggestion based on humanity, client's style, and environment.

香港是一片水泥丛林，处处都是高楼大厦，想要在街上找一些接近大自然的地方的确有点难，当你在一日的辛劳工作过后，回到家中只看着四面墙，或是往窗外看，一整片都是"楼景"，那种感觉令人多么难过！
所以此次作品取名为"城中原林"，目的是要将业主从每日的"水泥丛林"带回到一个被大自然包围的地方，利用材质的协调去塑造自然风。
环顾一室都是业主钟情的大地色系，以营造自然、简洁的风格，电视墙身也裱贴有石材质感的壁纸，以维持整体的一致。屋顶上原有的一条横梁，以木层架修饰，由内向外的线条延伸，将视线带到与户外相连的餐厅。

Living in Hong Kong is like Life in a concrete jungle, when you look up at all the skyscrapers and high rise buildings, it hard to get close to nature. This work named "Forest City". Designers use different type of materials to create Earthy and natural environment. Make you feel like you're walking into a relaxing place surrounded by beautiful nature.
Look around a room is the head of the household of the earth color, to create a natural and concise style. The TV wall body is mounted with a stone texture wallpaper in order to maintain the overall consensus on the original a beam. The modified layer with wood racks and line extends from inside to outside, to bring visual from outdoor to dining room.

Apartment Space 公寓空间

Apartment Space 公寓空间

淳净木沐
Pure Wood

项目地点：台湾省新北市新庄区思源路 680 号 19 楼 /Location : Building 19, NO.680 Siyuan Road, Xinzhuang District, Xinbei, Taiwan
项目面积：150 平方米 /Area : 150 m²
公司名称：金湛室内设计有限公司 /Organization Name : Goldesign Studio
设 计 师：凌志谟 /Designer : Ling Chih-Mo

IAI 设计优胜奖

凌志谟 Ling Chih-Mo
中国台湾 Taiwan, China

2007 年创立金湛空间设计研究室 Goldesign Studio, 致力于透过业主的内涵与现代风格反映出生活的空间时尚品位和文化层次，作品崇尚简约，线条比例充满美感。

透过创意延伸居住者的生活精神与内心期望，跳脱既定符号的框架与设限，逐一引导居住者从情绪酝酿至心境内化，通过界限之间的转折层次，触动个人深藏的记忆想念，寻求生活中种种美好的可能。

Ling Chih-mo established Jin Zhan Interior Design Co., Ltd. In 2007. He wants to show the taste of life space and culture according to the owner's characteristic and modern style. The feature of his work is simple, graceful lines.

Through the creative, extend the residents' spirit of the life and desire, jump out of limits and restrictions, guide habitant from brewing mood to internalization. By way of cross of different limits, touch people's deep memories, visit all sorts of beautiful in the life.

白色大理石、浅色木纹、不锈钢镜面、透明玻璃的使用，表现出纯净、明亮的现代北欧风格。
顶棚使用折纸原理修饰主要空间的大梁，顾虑到顶棚高度，不只是一个维度的不规则，连其他两个维度都有不同方向爬升与角度上的变化。从书房延伸到大门口，让客厅的氛围呈现简单、开阔和利落，三角形的灯光设计，活泼点亮室内，犹如静谧的湖水上轻点出涟漪的趣味。白色隔栅、大理石划分出客厅及厨房。而书房利用玻璃，做出独立的空间，即使在书房，也能看到客厅、餐厅，增加了家人之间的互动性。
整体流畅的动线，无阻隔墙开放式的概念，制造出视觉延伸，却也不失空间不同的功能与氛围。

The use of white color, white marble, light-color wood grain, stainless-steel mirror surface, and transparent glass shows the pure and bright modern North-Europe style. The ceiling modifies the main girder using the principle of paper folding. With consideration to the height of the ceiling, it's not only irregular in the X dimension, but also stretching to different directions and changing in different angles both in Y and Z dimension. From the study to the gate, the living room looks simple, open and tidy. The triangular lighting design makes a lively house, which looks like the ripples on a quiet lake. The living room and the kitchen are divided by the gondola made of white retainer and marble. The study is an independent place divided by the glass, so even when you are in the study, you can also see the living room and dining room, which improve the interaction between family members.
The integral and fluent line, along with the conception of open space without wall, creates a visual extension and keeps functions and atmosphere of different rooms.

Apartment Space 公寓空间

草根清境
Grassroots

项目地点：台湾省台北市北投区 62—1 号 15 楼 /Location : 15F,Building1, Beitou District, Taibei, Taiwan
项目面积：120 平方米 /Area : 120 m²
公司名称：金湛室内设计有限公司 /Organization Name : Goldesign Studio
设 计 师：凌志谟 /Designer : Ling Chih-Mo

IAI 设计优胜奖

此空间为度假用住宅，运用木头，米色和灰色大理石表现温和、内敛的氛围。在简单的线条分割中加入人文的元素，让空间中有休闲、放松的气氛却不显复杂。从客厅空间配置开始，使用木地板、暖色家具表现简约、沉稳的东方人文。进到餐厅，顶棚的格栅与不规则的灯光，质朴的木质餐桌与现代风格的餐椅搭配，在人文基调内添加了些许现代感。书房以黑色铁件制成的滑门区隔，门关起来时，可以当书房、客房；门打开时可以当起居室，以及眺望远方时的最佳地点。自然宽敞的开放空间，干净、利落的空间分割，使度假时家人聚会的功能增强，并且可以依照使用需求，营造出不同的情境感官享受。

The holiday house used wood and beigegrey marble to create a soft, modest atmosphere.A wooden floor and warm-color furniture give the living room a minimalistic and sedate oriental quality. The grille ceiling, irregular lighting and plain wood dining table, matched with modern dining chairs, add a modern sense to the dining room.The black iron sliding doors separate the study room from the rest, and can also serve as a guest room when closed, or a lounge when opened.With simple line partitions, the house is highly functional for family gatherings or for creating different sensational enjoyment according to the various needs.

Apartment Space 公寓空间

科技光感·时尚美学新面貌
Lighting Technology · Fashion Aesthetics New Look

项目地点：台湾省台北市敦化南路一段 /Location : Dunhua South Road, Taibei, Taiwan
项目面积：76平方米 /Area : 76 m²
公司名称：界阳 & 大司室内设计 /Organization Name : Jie Yang Interior Design
设 计 师：马健凯 /Designer : Ma Chien-Kai

IAI 设计优胜奖

马健凯　Ma Chien-kai
中国台湾　Taiwan, China

界阳和大司室内设计 设计总监；
毕业于中国科技大学建筑设计系。

设计风格
（界阳）黑白时尚前卫；
（大司）自然人文典雅。

Designer director of Interior Design of Jie Yang Interior Design;
Graduate from Department of Architectural Design, University of Science and Technology of China.

Style of design
(Jie Yang) Black and white, fashion, avant-garde ;
(Da Si) Natural, humanity, elegance.

近三十年老屋的现代新解，业主期待以与众不同的科技时尚感呈现，指定设计师采用经典的激光束屏风吸引入门视线，无定向的切割线条从顶棚漫游至客厅电视墙，围塑出100英寸背投影电视的惊人视效，并借由电浆玻璃可切换接口穿透度的特性，与灯光搭配出时尚多变的科技新貌。

马建凯设计师擅长的一体成型技术依旧是注目焦点，悬空打灯的床架线条越过内嵌光沟的结构柱体，顺势蜿蜒出立面的活性曲度，及至窗边连结卧榻与书桌；采用卷帘保护隐私的主卫浴也从浴柜线条开始，接续浴缸后转折延伸到半高墙面与门坎，无接缝的人造石线条变化，以大型艺术装置概念呈现，让美感与空间感并具。

A modern solution designed for old buildings above 30 years. For owner expected for an unusual sense of technology feels, he appointed laser byobu screened the doorway. The indirectly carves extended on the horizon wandering to the television wall and represented an extraordinary visual effect on 100" rear-projection TV. Plasma display panel glass exhibited the various fashion styles combined with lights for the quality of switchable penetration of interface.

The spotlight is constantly on the integrally formed technique, which is the field of Ma Chien-Kai. Lines of bed frame which suspended the light crossed the pillar construction which has light embedded in trench extended a stereoscopic curve to bed and desk interconnected near the window. The main bathroom provided privacy by a roller shower curtain started the lines with cabinets as well. The continuous line turned and extended from bathtub to the low wall and threshold. The alteration of line on seamless artificial stone is presented in the concept of large-scale installations in order to give consideration to aesthetic and perception.

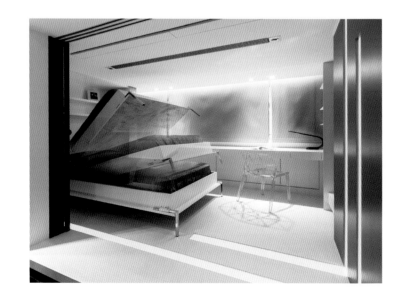

Apartment Space 公寓空间

景上苑庄邸
Garden View Of Yuanzhuang Manson

项目地点：台湾省台北市北投区西安街 229-1 号 5F / Location : 5F, NO.229-1 Xian Road, Beitou District, Taibei, Taiwan
项目面积：100 平方米 / Area : 100 m²
公司名称：大集国际室内装修设计工程有限公司 / Organization Name : Daji International Interior Decoration Engineering,Co., Ltd
设 计 师：高志豪 / Designer : Kao Chih Hao

IAI 设计优胜奖

高志豪　Kao Chih Hao
中国台湾　Taiwan, China

大集国际室内装修设计工程总监；
十余年的设计、工程经历；
国立台湾师范大学设计研究所；
中华大学建筑与都市计划系学士。

设计理念
从设计里，探究之于生活的必然性，经过模拟生活的功能运用及从平衡与比例概念里，延伸出空间的丰富和深度的层次美学，转换成情感和知性的融合，成就空间里最丰富的态度。

The director of Daji International Interior Decoration Eng,Co., Ltd ;
More than 10 years of design and project experience ;
Bachelor of Design institute of national Taiwan normal university ;
The Chinese university of architecture and urban planning department.

Design concept
In design ,they can explore the necessary in life.Through simulation life function and from the concept of balance and proportion,extend the space of rich and deep level of aesthetic, transformed it into the fusion of the emotional and intellectual, achieve the most abundant attitude in the space.

高志豪想从建筑的手法去强调空间内该有的虚实对比层次，在入口处设计了一道卡拉拉白色大理石墙。它是一道厨房与客厅的隔间。大理石墙叙述了一个庄重且永恒的质感。为了呼应与大理石墙同一轴线的电视墙，他设计了一个形如二维码的造型格栅，这个以线与面的交互对比关系，他想用这样的建筑元素活跃原本客厅与餐厅的气氛。在厨房与餐厅之间的立面，高志豪规划了主墙与双面玻璃推拉门，让木皮主墙面这样的立面连接到吧台的台面并连接到餐桌，强调出这一轴线，让轴线发挥出餐厅稳重且大气的氛围。大集的设计师们注重材料之间的对话，如大理石与铁件，铁件与木皮，刚性与柔性之间的对比关系。高志豪主张塑造大气的氛围，考虑到温润感，增加了木皮等元素。
从客、餐厅以黑白为主的彩度，打造出摩登的质地。而从卧室等私密空间规划里，大集的设计师们依照三个女主人心目中的梦想打造其空间。

The contrasting layers of space are emphasized through architectural techniques. At the entrance, a Bianco Carrara wall that gives the feeling of elegance and eternity,separates the kitchen from the living room. In contrast to the fullness of the marble wall, a two-dimensional barcode-like grid TV wall is placed in the same plane. The crisscrossing of lines and planes lifts the atmosphere inside the living room and dining area. As for the kitchen and the dining room, a wood veneer decorated main wall is in place with a double-paneled glass sliding door. Coupled with the alignment of the bar table and dining table, an axial line is formed to express a sense of stability and grandeur. The emphasis is placed on the relationship between materials,marble versus metal,or metal versus wood.It is also a contrasting relationship of rigidity versus flexibility versus flexibility.The overall effect of grandeur is softened with a gentle touch of wood elements.
An expression of modernism, the theme of black and white is evident throughout the living room and kitchen. On the other hand, the bedrooms are decorated based on the three hostesess.

Apartment Space 公寓空间

赋采
Poetry Colors

项目地点：台湾省台中市 /Location : Nantou, Taiwan
项目面积：189.38 平方米 /Area : 189.38 m²
公司名称：杨焕生设计事业有限公司 /Organization Name : Yhs Architecture& Interior Design Co.,Ltd.
设 计 师：杨焕生 /Designer : Jacksam Yang

IAI 设计优胜奖

杨焕生 Jacksam Yang
中国台湾 Taiwan, China

东海大学建筑硕士；
杨焕生建筑室内设计事务所主持人；
2015 年 现任亚洲大学室内设计学系讲师；
2013 年 美国《INTERIOR DESIGN》中文版杂志封面人物；
2005 年 成立杨焕生建筑室内设计事务所。

Master of Architecture Tunghai University ;
Design of Architecture Tunghai University ;
Design Director Yhs Architecture&Interior Design Co,Ltd.;
2015,Lecturer of School of Design ,Asia University;
2013,The cover people of INTERIOR DESIGN Magazine(America);
2005,Created Yhs Architecture &Interior Design Co.Ltd..

设计师们重新给予它像新生命绽放般的色彩，结合创作艺术与精致工艺，把它放在喧嚣都市的彼此交错坐落的城市光景中，也让这样的色彩巧妙融入生活。从玄关、客餐厅至厨房，长形的建筑空间，也是完全开放的尺度下，要让人不质疑这些各自独立的空间要如何并存，而起融合作用的，是将14幅连续且韵律感的晕染画作，模拟大山云雾的虚无缥渺，镶嵌于垂直面上，落实视角的想象，改变检视艺术的视角角度，实践内心期望的生活方式，一开一合之间创造出静态韵律与动态界面屏风，让连续性的延伸感蔓延全室，用弧形线条，如卷纸轴般的轻巧挂于顶棚上，饱满及圆润并攀延至墙面及柱体，使每一面视野都有自己的诗篇在流露，创造优雅、舒适的美好生活。

Designers re-bloom it like a new life of colors, combined with the art creation and fine craft; put it in the hustle and bustle city, located in the city staggered each circumstances, and also to let the colors vividly integrated into life.
From the entrance, dining room to the kitchen, long building space, under the completely open scale is to let people no doubt these separate spaces how to exist at the same moment. As for the fusion, integration of the role of the sky, is 14 continuous and rhythmic blooming paintings, simulation mountain clouds endlessly, inlay on the vertical domain, and implement perspective imagination, changing the perspective angle view of art, carry out inner expectations lifestyle. While opening and closing, it creates between static and dynamic interface screens rhythm, so that continuity extends to the whole room. With curved lines, such as the spindle-like light hanging on the ceiling, full and mellow and climbing extended to wall and cylinder, so that each side has its own vision of the Psalms in the outpouring, creating an elegant and comfortable better life.

Apartment Space 公寓空间

光影流转
Light Flow

项目地点：台湾省台中市国美馆特区绿园道第一排 /Location : Row 1, Jinguo Lvyuandao Parkway, Taichung, Taiwan
项目面积：387 平方米 /Area : 387 m²
公司名称：欣磐石建筑空间规划事务所 /Organization Name : Architecture & Space Planning Firm
设 计 师：罗仕哲 /Designer : Eric Luo

IAI 设计优胜奖

罗仕哲　Eric Luo
中国台湾　Taiwan, China

欣磐石建筑空间规划事务所设计总监罗仕哲先生，具有十年以上经验，对美学、艺术、建筑和精致工艺有独特的见解，结合工程团队专业工法和严苛监工，针对业主不同需求及多元风格，量身打造专属空间，营造泱泱大器，淬炼华丽风格，独创个人品位。

Design Director of Taiwan Architecture& Interiors &Space Planning Firm Mr.Luo with more than ten years' experience has unique views in aesthetics,art,architecture and craftsmanship. Eric Luo tries to consider unique taste into clients'demands ,meanwhile combining profession of team with strict supervisor,nd build dedicated space.

凝缩——玄关通道的端景是一面仅有一水平开口的大理石墙，视线在被阻挡与被开放之间，更加强了墙面之后光线的强度与公共空间宽度的感受。作为整个空间动线与光线路径的起始点，玄关赋予居住者一个可停顿与转折的可能，同时也是家与外界的过渡场所。

流动——光线与动线的自由度在这里是规划者的重点之一，结合本案优异的采光条件，在以不破坏空间连续性的前提下，由柜体、吧台作为空间的分割，具有穿透性的营造让空间感更加放大，因而能伴随着流畅的自由度。

Condensation——The side view of porch channel is a marble wall with only one horizontal opening, the vision is between blocked and opened, which strengthens the light intensity behind the wall and the spacious feel of public space. As the starting point of circulation and light path of entire space, porch gives occupants a possibility of standstill and turn, as well as the transition field of home and outside world.

Flowing——Here, the degree of freedom of circulation and light is one of key of planners, by combining with excellent lighting conditions in this case, under the precondition of not destroying the spatial continuity, the cabinet and bar are taken as the schism of space, which can amplify the space more with the creation of penetrability, thus it can along with the flowing freedom.

游走——在落地窗的底端,或是顶棚的水平线,规划者将消解大梁、包覆管线等工程上的难题结合视觉设计一并处理。水平线的生长与延伸,延伸视觉亦连结空间,创造出更具流畅的空间感。

Wandering——As for bottom of french windows, or horizontal line of ceiling, the planners will handle the visual design together with the problems in engineering, such as dispelling girders, coating pipelines. The growth and extension of horizontal line extend the visual and link the space to create a more smoother space.

Apartment Space 公寓空间

石宅
S Residence

项目地点：台湾省台北市万华区东园街 /Location : Dongyuan Road, Wanhua District, Taibei, Taiwan
项目面积：92.4 平方米 /Area : 92.4 m²
公司名称：晨室空间设计有限公司 /Organization Name : Chen Interior Design
设 计 师：陈正晨 /Designer : Chen Cheng-Chen

IAI 设计优胜奖

光线透视——以光线为主轴穿透整个空间让结构上成一直线，让主卧、起居室的光线无阻碍的渲染空间。
材质的延伸——以木作墙面去串连每个空间，达到空间基调的一致性，餐厅、和室、客厅面向安排在同一轴线上，表现出小空间配置性的各种可能。
空间的开放——以无隔间的概念将不同功能的空间整合在一地，也展现了设计师一贯将重心放在活动性、自主性规划的风格，不必多预留空间，利用开放和连贯的功能，衍生出场所的焦点。

Penetrating Ligh——Light is the center of design penetrating the entire space to create a linear prospect in which the light disperses without barriers throughout the main bedroom as well as other living spaces.
Extension of materials——Wooden walls connect each room and space to achieve a sense of spacial uniformity. The restaurant, living room and the tatami house are aligned to showcase designer's creativity in the overall layout of smaller spaces.
Openness of space——The concept of sans partition brings together spaces of different functions and displays the designer's usual emphasis on mobility and autonomy. Spaces are not reserved or secluded, but overt and continuous, one linking another to navigate the spacial focus.

陈正晨 Chen Cheng-Chen
中国台湾 Taiwan, China

2009 年 晨室空间设计有限公司成立 / 负责人
2005 年 晨室空间设计工作室成立 / 负责人
2003 年 唐狮室内设计工程有限公司
　　　　专业经理及大陆事业部负责人
2001—2003 年 中国科技大学 / 建筑系
2000 年 大三亿营造 / 建筑工程师
1999 年 德宝营造 / 建筑工程师
1998 年 台湾大学城市与乡村研究所 / 宜兰工作室
1994—1999 年 兰阳技术学院 / 建筑系

2009,Chen Interior Design Co., Ltd.,Charger
2005,Chen Interior Design Studio,Charger
2003,Tang lion interior design engineering co., LTD, Professional Manager,Person in charge in the mainland division.
2001—2003,University of Science and Technology of China,Department of Architecture
2000,Dasanyi Business,Engineer
1999,Debao building,Engineer
1998,Institute of urban and rural, national Taiwan university,LanYi studio
1994—1999,Lan Yang Institute of Technology, Department of Architecture

Apartment Space 公寓空间

当代艺术宅
Contemporary Art Residence

项目地点：台湾省桃园市 / Location : Taoyuan, Taiwan
项目面积：178平方米 / Area : 178 m²
公司名称：青埕空间整合设计 / Organization Name : Clearspace Architecture and Interior Design
设 计 师：郭侠邑 / Designer : Ryan Kuo

IAI 设计优胜奖

郭侠邑 Ryan Kuo
中国台湾 Taiwan, China

青埕空间整合设计有限公司负责人主持设计师
青埕空间整合设计、建筑与人文空间研究。

设计特点：生态自然、当代艺术、人文时尚。
一直以来把绿建筑概念，试着融入到各层面的空间设计当中。
服务内容着重于建筑整合、空间设计、私宅订制、生态绿建筑、节能住宅和养身会所，以及商业空间、医疗办公、餐饮酒店、家饰美学和视觉形象整合规划设计。

Director of Clearspace Architecture and Interior Design;
Clearspace Architecture and Research on Humanistic Space;
Design Features,ecologi calnature,contemporary arts and human fashions.ClearSpace Architecture and Interior Design with the concept of green building has tried to integrate in diverse space design. It focuses on the design of architectural integration,space design,private residence,green building,energy-efficient residence,health club,commercial space, medical office,restaurant and hotel ,decoration aesthetics and visual brand design.

白色能与世界上所有华丽的色彩相搭配，它能延续近三分之一广度，业主可自主地享受在放大的空间漫步。

本案白净通透宛如美术馆，使得空间的本质、生活的本身及大大小小的艺术品都尽兴地展演自己。艺术才是生活中的主角，艺术即生活，生活即艺术。打造通透白净、平整的格局，再引入自然光线，彷佛重现人与环境关系的美术馆，让人在室内感到从容自在。通透的格局，可以是艺廊，也可以开派对。家具的陈设多以活动式为主，可以随不同需求进行调整，以达到空间使用的最大效益。

The color white captures within itself all of the gorgeous colors in the world. Able to expand a sense of space, the color makes the space seem one-third larger. This allows one to feel as if they were autonomously enjoying an enlarged space. The goal is to be white but penetrable, echoing a museum. The nature of space, life itself and the art therein together create a sense of performance. Art emerges as the protagonist in life. Art is life,life is Art. Created as LOFT style – white, clean, using a smooth pattern with plenty of sunlight. It reproduces the relationship between people and the environment found in a museum. It is calm and free.
The transparent design is good for either a gallery or for a party space. The furniture display emphasizes mobility, allowing for adjustment based on different needs. The goal is to allow for the maximum effectiveness of space usage.

Apartment Space 公寓空间

圆舞
Rounded House

项目地点：台湾省台北市四天祥路 /Location : Si tianxiang Road, Taibei, Taiwan
项目面积：33 平方米 /Area : 33 m²
公司名称：欣琦翊设计有限公司 /Organization Name : C.H.I. Design
设 计 师：林真琦、何欣 /Designer : Lin Chen-chi, Ho Hsin

IAI 设计优胜奖

C.H.I. Design Studio

何欣 Ho Hsin　林真琦 Lin Chen-Chi
中国台湾　Taiwan, China

2008 年　欣琦翊设计有限公司项目设计主持；
2008 年　中原大学室内设计系兼任讲师；
2003—2008 年　太一国际设计项目设计。

Since 2008　Director of project design of CHI Design Studio;
Since 2008　Adjunct Lecturer at Department of Interior Design of Chung Yuan Christian University;
2003—2008　Project designer of Taiyi International Design Company .

室内整体面积为33平方米的小空间既是限制，同时也是整体概念的基础发想，以圆角作为概念出发，企图让生活在其中的人，能够因为圆角的滑润修饰掉空间原有的狭隘感，以及利用挑高的特性加强层高的变化，创造出复层空间中不同视角的经验，同时区分而成的私密和公共空间，并且以垂直的轴线去规划储藏功能 。
柜体设计成活动滑轮式可作收纳同时也可当桌椅使用，让即便是小空间也有宽敞而干净的动线。最后将白色涂抹在整体空间之中，搭配底面黑色让视角随空间流动着。
整体空间在材质的选择上以白色为主要材料，以最简化性的单一材质减少材质的消耗，来实现创意的环保概念。

Owing that the interior space is confined to 33 square meters, which is considerably small , the key concept of design is derived from curves. The residents are provided with refreshing experience as they live in a multilayer space. On one hand, the intrinsic narrowness is substantially alleviated by smooth curves, while the variation of height is greatly intensified by its characteristic of high ceiling. On the other hand, in either private or public areas, the storage space is enlarged by vertical axes, whilst its curves are polished. Concerning the cabinets, they can be utilized as chairs or storage space so as to arrange the limited interior space by wide and smooth routes. Lastly, the viewers' perspectives are constantly mobilized since the interior space is whitewashed and its bottom part is painted in black.
In the materials, choosing the simplicity with white ones to be the main character, reaching the concept of life-loving people a feeling to nature a feeling in a green environmental protection.

佛山东怡美居设计
Foshan Dong Yi Beauty in Design

项目地点：广东省佛山市顺德 /Location : Shunde District, Foshan, Guangdong
项目面积：220 平方米 /Area : 220 m²
公司名称：Max 马思中国（佛山室内设计）/Organization Name : Max China, Foshan Interiors
设 计 师：谢法新 /Designer : Xie Faxin

IAI 设计优胜奖

谢法新　Xie FTaxin
中国　China

米兰理工大学国际室内设计学院在读硕士；
国家二级建造师；
IAID 十大新锐设计师。

Postgraduates in school in school of international interior design of MilanPolytechnic University;
National secondary constructor;
The Top Ten Young Designers in IAID.

这是一间位于佛山顺德区的旧公寓，楼龄超过 15 年，漏水，管线老旧等问题都已经浮现出来了。隔间、卫生间、厨房等空间也不符合年轻业主的现有需求，设计师决定进行整体改造。

设计师将餐厨区安排成入门的第一个区域，并呼应电视背景墙增设一个吧台，将客、餐厅做开放式安排，完成整个公共区的骨架。开放式设计不仅让客餐厅动线更顺畅，也令视线能自在游走，对于放大空间视觉效果颇有助益。由于之前房屋的设计不合理，使非常热爱运动的户主没有一个舒适的区域健身。由此，我们把健身区域安排在户外花园旁边，希望业主可以一边运动一边感受大自然的气息。

This is an old apartment in the Foshan Shunde District, the building is over 15 years old, Water Leakage ,pipeline and so on, old problems have been emerged. Compartment, toilet, kitchen space is not to meet the existing needs of young householder,designers decided to carry out the overall transformation. Designers will be the first region kitchen area arranged in the entry, and echoed the television background wall is additionally provided with a bar, restaurant, passengers will be open arrangement, complete the whole skeleton of public area.Open design not only to the customer, the restaurant line more smoothly, also make the line of sight can be free, for the enlarged space visual effects helpful.Because of the design before the housing unreasonable, the very love sports without a comfortable area fitness. Thus designers have arrangements in gym, outdoor garden next to the head of household, hope can while movement while feel the breath of nature.

Model House Space
样板房空间

The A3 Model House of No.8 Building in Poly Sanshan New City Seattle Project Material ——Give a Salute to Mondrian	240
Greentown YuYuan Villa	242
Shanghai Vanke Wujie Fang	244

240	保利三山新城西雅图 8 栋 A3 样板房 ——向蒙德里安致敬
242	绿城御园·法式合院别墅样板房
244	上海万科伍玠坊

保利三山新城西雅图 8 栋 A3 样板房——向蒙德里安致敬
The A3 Model House of No.8 Building in Poly Sanshan New City Seattle Project Material——Give a salute to Mondrian

项目地点：广东省佛山市 /Location : Foshan, Guangdong
项目面积：58 平方米 /Area : 58 m²
公司名称：广州道胜装饰设计有限公司 /Organization Name : Guangzhou DS Design
设 计 师：何永明 /Designer : Tony He

IAI 设计优胜奖

何永明 Tony Ho
中国 China

2008 年 华南师范大学室内设计系客座讲师；
2007 年 广东工程技术学院客座讲师；
2005 年 中国室内设计师协会注册设计师；
2005 年 成立广州道胜装饰设计有限公司；
2003 年 成立何永明设计师事务所（主要从事建筑和室内设计）；
2001 年 任职广东雅风设计事务所（设计总监）；
1999 年 任职广东省广告公司（工艺美术师）。

2008 Guest Lecturer of South China Normal University;
2007 Guest Lecturer of Guangdong Engineering Polytechnic;
2005 Certified Designer of CIDA;
2005 Establish DS Design Company ;
2003 Establish He Yongming Designer Firm;(work mainly on Architecture and Interior Design);
2001 Work at Guangdong Yafeng Design Firm (Design Director);
1999 Work at an Advertising Company at Guangdong (Industrial Artist);

本案运用"蒙德里安红黄蓝的直线美"为灵感，采用大小不等的红、黄、蓝创造出强烈的色彩对比。画面由长短不同的水平线和垂直线分割成大小不一的相互穿插的正方形和长方形，并以粗黑的交叉线将他们分开，并运用原木家具来增加空间的自然与和谐。原色与黑、白、灰的对比，排列就像音符旋律中的变化。室内的饰品、挂画、地毯等都细腻的延续着蒙德里安元素，色彩大胆跳跃。在儿童房中，有趣的墙贴和一地的玩具，诠释出儿童开朗活泼的性格。主卧在色彩丰富的墙面和地毯上，用素雅的浅灰色来中和过渡，让空间丰富的同时也能稳重而不浮躁。

This plan uses "the Liner Beauty of Mondrian Red Yellow Blue" as the element. It creates strong color contrast and stable balance with unequal sized red, yellow and blue.The picture is divided in to squares and rectangles of different sizes with different horizontal and vertical lines. The bold intersection line separates them and insert rectangles around the squares.Those primary colors and the contrast of black ,and the use of natural wood in furniture adds the nature and harmony in space.Gray and white are arranged like the note changes in melody. Accessory, picture and carpet continues the element of Mondrian and boldcolor. In the children's bedroom, interesting wall paper and toys on the ground explains the children's cheerful personality.The colorful use on the wall and carpet in the master bedroom is blended by the elegant light gray ,which makes the abundant space stable but not blundering.

绿城御园·法式合院别墅样板房
Greentown YuYuan Villa

项目地点：浙江省湖州市仁皇山新区青铜路1739号法式合院 /Location : No.1739, Qingtong Road, Renhuangshan New Area, Huzhou, Zhejiang
项目面积：607平方米 /Area : 607 m²
公司名称：上海唯思室内设计有限公司 /Organization Name : Shanghai Wise Interior Design Co.,Ltd.
设 计 师：王玮，王征宇 /Designer : Wang Wei, Wang Zhengyu

IAI 设计优胜奖

上海唯思室内设计有限公司

上海唯思室内设计有限公司
Shanghai Wise Interior Design Co.,Ltd.
中国 China

上海唯思室内设计有限公司是集住宅空间设计、办公空间设计、商业空间设计、软装高端订制、海外项目一站式订制等为一身的室内设计专业服务机构。
以"睿思谨行、唯精至求"为设计准绳，凭借专业、敬业的精英团队，系统性地构建了从软硬装的前期策划、概念设计到后期深化设计与实施的整体解决方案。

"Wise Interior Design"offers diverse services including living space, office space, commercial space, decoration custom, overseas projects and one-stop customized services,etc.
With design principle of"wisdom and caution,the pursuit of refinement",systematically builds the solution from prophase plan of decoration,conceptual design to the final detailed design.

上有天堂，下有苏杭，湖州在天堂中央。
门有流水，院有庭芳，逸韵在御园品赏。
这是面积为607平方米的一套独栋住宅，依山而筑，处在缓坡之上，拥有绝佳的私享景观。户型为法式合院设计，中央四方的挑空庭院令阳光透过玻璃穿梭在房间每一个角落，与建筑的结构融为一体，使得空间呈现雕刻般剪影的变化，形态和材质的微妙结合使得住宅的气质既尊贵又含蓄。
此作品是与绿城集团合作的一个样板房项目，位于浙江省湖州市，是一套独栋豪宅。本案的设计在Art Deco风格的基础上，融入了一股中式元素的清新风，将低调内敛的贵族气质与住宅建筑设计宽敞大气相融合，创造出一个具有"凝神具气、气脉相连"的灵性空间。

Through Art Deco style, the designer brings the owners a sense of unlimited luxury, and meanwhile makes them feel profound and classical temperament from the orient.The appeal of this design is how to maintain and promote Chinese traditional style under such background as styles of Europe and America are popular nowadays. The house is located in Huzhou, Zhejiang, a noted historic and cultural city, and the design theme is developed by introduction of elements of South Yangzi River. The designer combines the local materials and traditional handicrafts of most characteristics, chooses four seasons as the topic, and makes all these cultural elements integrate into Art Deco to achieve a natural and perfect effect.

挑高的客厅以四季中特有的花卉作为背景屏风，在水晶吊灯的绚丽光芒下，奢华中带有一丝清新的中国风。整个样板房是以四季为装饰线索串连了整个空间，并且配合江南特有工艺特制而成，桃花图案是通过苏绣制作的除了刺绣，另一种中国传统工艺——漆画也被运用上了，一幅梵高的《杏花》色彩明丽、工艺精湛，将室内装点的雅致而抚媚；融合江南一带特有的文化元素订制成的装饰橱柜将道不尽的江南文化凝于室内；毗邻的餐厅瑰丽堂皇，一盏螺旋形水晶灯自顶棚徐徐垂下，点点灯光洒落在八角形的坐椅上，焕发矜贵和谐之感。主人卧室内，沉稳大气的蓝色花纹地毯为房间注入温馨自然的格调。床背墙以植物图案扣布饰面，配合镶嵌水晶柱和白贝母的造型、超大比例的白色大床，散发和谐雅致的韵味。

The sitting room with tall ceiling is in the background screen of four seasons of flowers,under the gorgeous light of crystal chandeliers,luxury with a hint of fresh Chinese wind.The example room is decorated in the four seasons series the whole space,and with Jiangnan special process.The peach flower pattern is made by suzhou embroidery ,also with the Chinese traditional craft lacquer painting.A pair of Van Gogh "Prunus"is in bright color,exquisite workmanship ,decorate the indoor with elegant and charming,fusion in jiangnan area culture making the fixed adornment cabinet in the interior.Next to the magnificent and grand restaurant,a spiral crystal light from smallpox down slowly ,and the light on the octagonal chair ,making a harmonious and noble atmosphere.In the master bedroom,composed of atmospheric blue patterned carpet for the room into sweet and natural style. The bed wall cloth facing with plant design,with embedded column crystal and white of the modeling,super large proportion bed ,all in a harmonious and elegant charm.

Model House Space 样板房空间

上海万科伍玠坊
Shanghai Vanke Wujie Fang

项目地点：上海市浦东区东明路杨南路 / Location : Dongming Road, Pudong District, Shanghai
项目面积：320 平方米 / Area : 320 m²
公司名称：上海涞澳装饰设计有限公司 / Organization Name : IADC Designevrs. Ltd.
设 计 师：潘及 / Designer : Eva Pan

IAI 设计优胜奖

潘及 Eva Pan
中国 China

潘及 IADC 涞澳设计公司设计总监；
米兰理工大学室内设计管理学专业硕士学位；
其设计项目分布在中国许多重要城市，透过整合建筑、室内设计、视觉图像和室内布置，每一次新作皆创造出独特的感官魅力与欢愉的空间氛围；
与万科集团、凯德置地、吉宝置业、金地集团和旭辉集团均有长期合作，并创造了多个经典作品。

Design Director in IADC Lai Design Company in Australia；
Professional Master's Degree in Interior Design Management in Milan Polytechnic University；
Its design project distributed in many important cities in China, through the integration of visual images and indoor decorate interior design,；
every new works are creating a unique charm and joy of space atmosphere of the senses；
Long-term cooperation with Vanke,Capitaland, Keppel Land,Gemdale Corporation and Xuhui group, and creating a number of classic works.

在这个案例中，设计师运用着东方元素，通过家具、面料、饰品来表现．同时也利用了一些西方的方式和当代的表现手法。颜色的跳跃、材质的多元化、使得空间游走在东方文化和西方当代生活中，呈现丰富且具有内涵的气质。
在设计过程中，还考虑了业主的背景，以他的特点作为设计脉络，男主人以一个金融投资者为人物定位，留学归来，身受西方的教育，女主人则是一个热爱艺术的全职妈妈，家有一儿一女，生活非常美满。他们有着同样的爱好，就是收藏画和摄影作品。女主人对艺术的品位，和对生活的热爱在空间中体现得淋漓尽致，正是这样的人物背景，让我们的空间中弥漫着东西混合的特殊韵味。其中我们还运用了 Hermes 的主题，来延续他们的爱好，比如骑马等，价值不菲的 Hermes 马鞍、餐具、毯子等，让空间增色，来自意大利 Armanicasa、Flexform、Minotti 等的家具，都是世界高端的家具品牌。

In this case, the oriental element is applied on the furniture, fabric, adornment and etc. Meanwhile, contemporary and western expression means were skillfully added into the design. The rich color and diversified materials made Eastern and Western cultures collide with each other, and express the meaning of temperament.
The character positioning is a key consideration during the design. The host is positioned as a man, who works in a financial investment firm, has a foreign study background. The hostess is envisioned as a stay-at-home wife, who loves art. And they both loved to collect works of art and photographs. Such a design and character positioning makes the whole space present a mixture of the eastern and western style. On home accessories selection, Hermes's saddle, tableware and blanket were associated with master interest Riding.

Building Space
建筑空间

Koyasan Guest House	248
Dig in the Sky	250
New Kyoto Town House	252
Wood TeK HQ	253
Olgiata Shopping Plaza	256
Olgiata Sporting Club	258
Villae Minimae	260
Semi-Mountain Architecture	262
CR Shimei Bay International Marina Club	266
UNASUR General Secretary Headquarters	268
Impossible Space Beijing	272
Times Museum	274
Mountain House——Tourist Center of Shanchuan, Anji	276
Hengqin Yangchang Museum	278
The Impression of Wu Long Primitive Town	280

248	高野山宾馆
250	天井
252	新京都宅
253	Wood Tek 总部
256	奥杰塔购物广场
258	奥杰塔运动俱乐部
260	小别墅
262	半山建筑
266	华润石梅湾国际游艇会会所
268	UNASUR 秘书总部
272	北京不可能空间
274	时代美术馆
276	竹石山房——安吉山川乡游客中心设计
278	横琴杨昶博物馆
280	印象武隆小镇规划设计

Building Space 建筑空间

高野山宾馆
Koyasan Guest House

项目地点：日本京都上京区 /Location : Kamigyo-ku, Kyoto, Japan
项目面积：137.26 平方米 /Area : 137.26 m²
公司名称：Alphaville 建筑有限公司 /Organization Name : Alphaville Architects Co., Ltd.
设 计 师：坂口健太郎，山本麻子 /Designer : Kentaro Takeguchi, Asako Yamamoto

IAI 创意大奖

Alphaville 建筑有限公司
Alphaville Architects Co., Ltd.

日本 Japan

坂口健太郎
Kentaro Takeguchi

2012—至今　大阪产业大学兼职教授，神户大学讲师；
1998 年　与 Asako Yamamoto 合作建立了 ALPHAVILLE Architects 事务所；
1998 年　京都大学建筑学系硕士毕业；
1995 年　英国建筑协会建筑学院学习；
1994 年　毕业于京都大学工程学院建筑学系。

2012　Adjunct professor at Osaka Sangyo University Lecturer at Kobe University;
1998　Established ALPHAVILLE Architects with Asako Yamamoto;
1998　Graduated from master course, School of architecture, Kyoto University;
1995　Studied in Architectural Association School of Architecture, U.K in Diploma Unit 5, Foreign Office Architects;
1994　Graduated from School of Architecture, Faculty of Engineering, Kyoto University.

山本麻子
Asako Yamamoto

2012—至今　滋贺县立大学讲师；
1998 年　与 Asako Yamamoto 合作建立了 ALPHAVILLE Architects 事务所；
1998 年　京都大学建筑学系硕士毕业；
1995 年　法国巴黎建筑学院学习；
1994 年　毕业于京都大学工程学院建筑学系。

1994　Graduated from School of Architecture, Faculty of Engineering, Kyoto University;
1995　Studied in l'Ecole d'Architecture de Paris, La Villette, France;
1997　Graduated from master course, school of architecture, Kyoto University;
1997　Worked at Riken Yamamoto & Field Shop;
1998　Established ALPHAVILLE Architects with Kentaro Takeguchi;
2010　Lecturer at University of Shiga Prefecture.

设计师设计了一个新的客房给在高野山的全世界的年轻人。真言宗主要的庙宇是在 1200 年以前建立的，是世界文化遗产。它是隐密性很好的日本"胶囊式"旅馆和交流十分活跃的宿舍的混合体。每一个独立房间都直接面对着一个大厅，所以可以选择与其他客人的合适距离来确保自己的私人空间。对于薄的木质结构的选择，是因为每根柱子相对较轻，环境设施更容易维护。并且这种简单的组合使得空间不仅能够让该客房的主人，而且同样也能让客人在这里住很长的一段时间。

Designers designed a new guesthouse for young people from all over the world at Koya-san, the head temple of the Shingon sect founded 1200 years ago, the UNESCO world heritage. It is a mixture of Japanese capsule type hotel in which the privacy is well protected, and dormitory in which the communication among the guests is active. Each single room directly faces a hall so that you can chose proper distance with other guests ensuring the privacy. Selecting of thin wooden structure, resulting that the burden load per one pillar is relatively light, visibility of environmental facilities for easier maintenance and the simple composition of the space allow not only owner of this guest house but also guests to maintain, modify and keep on using this architecture for the long time.

天井
Dig in the Sky

项目地点：日本大阪 / Location: Osaka Prefecture, Japan
项目面积：152 平方米 / Area: 152 m²
公司名称：Alphaville 建筑有限公司 / Organization Name: Alphaville Architects Co., Ltd.
设 计 师：坂口健太郎、山本麻子 / Designer: Kentaro Takeguchi, Asako Yamamoto

IAI 设计优胜奖

在大阪的中心，设计师设计了一个由三个大楼和两个天井组成的住宅，以便这个狭小区域能向临近的三个方向的房子借光。由于每栋房子每一层都是一个房间的大小，设计师决定用管状的走廊和楼梯来连接房间并穿过天井。每个房间都不是用门隔开的，取而代之的是长长的入口。它们连接得并不紧密，但是能够获得适宜的居所距离感。

At the center of Osaka, designers designed a residence that is composed of three buildings sandwiching two courtyards, so as to bring light to a narrow site surrounded by neighboring houses on three sides. As each building will hold one-room-sized space on each floor, designers decided to connect rooms with tube-shaped corridor and staircase that cross the courtyards. Every room is not divided by doors, but instead is linked by long stretching-shaped entrances. They are loosely connected, but are able to gain appropriate sense of distance for habitation.

新京都宅
New Kyoto Town House

项目地点：日本高野山 /Location : Koyasan, Japan
项目面积：78.68 平方米 /Area : 78.68 m²
公司名称：Alphaville 建筑有限公司 /Organization Name : Alphaville Architects Co., Ltd.
设 计 师：坂口健太郎、山本麻子 /Designer : Kentaro Takeguchi, Asako Yamamoto

IAI 设计优胜奖

这是一个处于京都中心的一个狭窄地点的住所，京都曾经是日本的首都。这片区域与传统的木制镇公所排列在一起。继承了镇公所的优点，设计师意图克服它们的缺点并创造一个舒适的享乐空间。这个房子最有个性的特点是多面的隔板墙形式。它们不是靠直觉制造出来的，而是依据逻辑概念和多功能的设计思维。

This is a residential house located on a narrow site in the centre of Kyoto, the old capital of Japan. The area is lined with traditional wooden townhouses. While inheriting the advantages of townhouses, designers intended to overcome their drawbacks and create a comfortable and enjoyable space. The most characteristic feature of this house is the polyhedral form of the partition walls. They are not made by intuition but are based on logical concepts and perform multiple functions.

Wood TEK 总部
Wood Tek HQ

项目地点：台湾省台中市西屯区 /Location : Situn District, Taichung, Taiwan
项目面积：518.07 平方米 /Area : 518.07 m²
公司名称：考工记工程顾问有限公司 /Organization Name : Origin Architects & Planners
设 计 师：洪育成 /Designer : Yu Cheng Hung

IAI 杰出设计大奖　IAI 年度最佳设计机构

考工记工程顾问有限公司
Origin Architects & Planners

设计核心价值
诗意建筑　健康建筑　永续设计
公司创立时，取"考工记"为名以自许。期待在建筑的"本质"探讨，寻求建筑的诗意境界，并在这瞬息万变的年代，对未来建筑提出前瞻性的看法。

"身心灵"整体健康是现代人追求的生活型态，"考工记"的核心价值之一在于创造健康的建筑，提供使用者在工作、休闲、居住健康的环境。不只是室内健康绿建材的运用，更把"阳光、空气、水"带入生活空间之中。使居住者享有健康的居住空间。

"永续设计"是考工记的另一核心价值。如何运用绿色建材，创造节能减碳的优质生活空间是考工记的专长。考工记的设计，由建筑设计、室内设计、景观设计到家具设计，都将"永续设计"的核心价值发挥到极致。

The core value of design
Poetic Architecture、Healthy Architecture、Sustainable Design:
Origin Architects & Planners has followed the records and ideas in Kao Gong Ji since the company was established. They look forword to exploring the essence of architecture, seeking the poetic boundary of architecture, and presenting some forward-looking statements on future architecture of this rapidly changing era.
"Body-Mind-Spirit":The overall health is an idea lifestyle that modern people pursuit. Origin Architects & Planners creates "healthy houses"and offers clients healthier working , leisure and living environment as one of its core values.Not only the use of interior green building materials, but also bring in "sunshine", "air", "water" into the living space, which makes the habitants enjoy a healthy living space.
"Sustainable Design" is another core value of Origin Architects & Planners. How to use green building materials and create energy conservation and low carbon life space is Origin Architects & Planners's expertise. From the architectural design, interior design, landscape design to furniture design, Origin Architects & Planners has fully showcased its core value of sustainable design.

WoodTek总部大楼坐落在筏子溪旁，眺望着来往的高铁，是台湾第一栋CLT（Cross Laminated Timber）材料的建筑。设计师希望这栋建筑不仅在外观上成为地标，也能作为一个象征亚洲绿建筑发展的里程碑。因所使用的木材都是人工林的经济树种，隔热和防火功能优异。从生态、节能和舒适安全方面来考量，都占有优势，而且可往中高层建筑发展。
结构系统是此项工程中的另一个挑战，我们试图打破 CLT 建筑多为"盒状"的刻板印象。造型外观上，建筑的表达如同雕塑——由实墙量体与透明玻璃梯间的实虚组合。在建筑物一侧，光线穿透过梯间进入室内创造出戏剧性的空间效果。

Overlooking the Farze Crook by the high speed train rail, Wood Tek headquarters building is the first ever CLT building in Taiwan.Designers wanted this architecture to be a landmark not just in it's physical appearance but also as a symbolic approach of Green Architecture development in Asia.
Structure system is another challenge for this project. Designers made this building step forward instead of back to express the advantage of using panels as walls and floors. Designers also tried to break the stereotype of " boxlike " CLT construction image. In it's exterior shape, architectural expression is articulated by the combination of solid walls and transparent glass staircase. In side of the buildings, light penetrating through this staircase casts a dramatic space effect.

Building Space 建筑空间

奥杰塔购物广场
Olgiata Shopping Plaza

项目地点：意大利罗马卡西亚大道 /Location : Via Cassia, Rome, Italy
项目面积：商店 2300 平方米，公园 2000 平方米 /Area : 2300m² Shop+2000m² Parking
公司名称：LAD 建筑和设计工作室 /Organization Name : LADs.r.l
设 计 师：费朗西斯科·那拨利塔诺、西蒙·拉奈罗 /Designers : Franceco Napolitano, Simone Lanaro

IAI 最佳设计机构

LAD 建筑和设计工作室
LAD s.r.l.
意大利 Italy

LAD 代表 "Architecture and Design Workshop" 建筑和设计工作室，涉及每一个设计过程的一部分。它成立于 2006 年，是由西蒙·拉奈罗和费朗西斯科·那拨利塔诺在意大利罗马建立的。
2010 年，成为意大利 40 位青年建筑之一，并在 2010 年上海世博会上，意大利展区展示他们的作品。2011 年，LAD 被 UTET 选刊在 GIARCH 上，这是大多数意大利年轻建筑师的参阅书籍。

LAD stands for "Architecture and Design Workshop" and is meant to involve every part of the design process. It was established in 2006 by Simone Lanaro and Francesco Napolitano and it is based in Rome, Italy. In 2010 ,LAD was one of the 40 young Italian Architecture offices ,exposing their work in the Italian Pavillon at Shanghai Expo 2010. In 2011 ,LAD was selected by UTET for the publication of GIARCH, a volume dedicated to young Italian architects.

要解决当今社会最具代表性的问题，就是人们为了社交与交流而在商店橱窗里消耗的时间。这个项目反映出的原则基于这些结构需要城市环境和郊区，起点就是减少体积带来的影响。我们寻找的形式不是减少有机而是相反地还原一个新的形象，同时给建筑物的表面带来生机。棱锥台的运用，人为的迹象和自然形成的土斜坡，营造了幻影控制的感觉和建筑与环境的和谐，同时也体现了历史的延续性。新楼上锌钛板的屋顶给与人们交流机会。如果建筑可以用言语来形容，则是干预和军事化的建筑，紧凑而必要。

In addressing one of the issues more representative of society, in the present, the consumption of and free time that has elapsed between the shop windows become points of election for the exchange and the sociability. The project reflects the principle on the consequences that the inclusion of these structures entails on urban contexts and suburban in which are inserted. Starting point is precisely the reduction of the impact that the volume provided by indexes could have against a context to strong naturalistic vocation, if declined canonically. The search for forms that however did not try the mimesis organic but that on the contrary return a new image and at the same time known, gave life to the surfaces of the building. The artificialisation of this truncated pyramid, tangible sign of the man while mindful of the natural slopes of the soil, that makes the new offers as an apparition controlled, dialectic synthesis between architecture and the environment, also understood as historical continuum. Mastaba as protecting a sacredness below, the coverage in titanium-zinc of the new building houses the places members to exchange and to local trade. If the architectures can be compared with the words, then the etymology of this intervention and military architecture, aerodynamics, compact and essential.

奥杰塔运动俱乐部
Olgiata Sporting Club

项目地点：意大利罗马 /Location : Rome, Italy
项目面积：7 500 平方米 /Area : 7 500 m²
公司名称：LAD 建筑和设计工作室 /Organization Name : LADs.r.l
设 计 师：费朗西斯科·那拨利塔诺、西蒙·拉奈罗 /Designers : Franceco Napolitano, Simone Lanaro

IAI 最爱佳设计机构

奥杰塔运动俱乐部位于意大利罗马边境的一个花园城市的森林中。楼房由三个不同但连在一起的楼阁组成，中间的一个是行政中心，它有一个入口，这个入口也是主楼梯和主连接处所在的地方。穿过这个入口，参观者能够到达低一点的衣帽间和另外两个楼阁。南大楼有体育馆，北大楼有游泳池。通往不同区域的路是完全分开的。室外空间有足球场、网球场和沙滩排球场。飞回棒（回飞棒，一种弯曲的投掷棍棒。）形状的梁是用胶合层木材质，由 Holzbau 在意大利设计制造。屋顶采用锌钛合金制作的，由 Rheinzink 制作。

The Olgiata Sporting Club is located inside the woods of a garden city at the border of Rome, Italy. The building is made up of three different pavilions connected to each other. The central one is the administrative center and contains the entrance, where the main staircase and the connections are. Through it the visitor can access the locker rooms at the lower level and then the two pavilions. South pavilion contains gyms, North pavilion contains swimming pools. The paths to reach the different areas are strictly separated. Outdoor space accommodates football fields, tennis courts and a beach-volley court. The boomerang shaped beams are made up of glulam, designed and produced by Holzbau in Italy. The roof is in zink-titanium, produced by Rheinzink.

Building Space 建筑空间

小别墅
Villae Minimae

公司名称：LAD 建筑和设计工作室 /Organization Name : LADs.r.l
设 计 师：费朗西斯科·那拨利塔诺、西蒙·拉奈罗 /Designers : Franceco Napolitano , Simone Lanaro

IAI 最佳设计机构

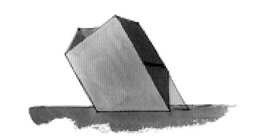

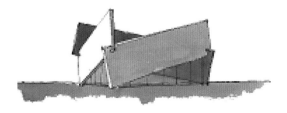

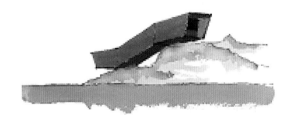

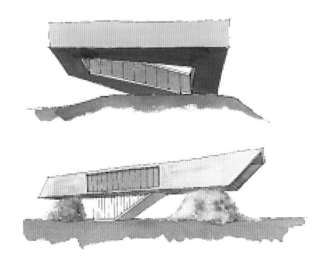

"Villae Minimae"是一个有五个独立家庭别墅组成的集合。它是以 1/200 的规模发展的。在所有的五个别墅中，项目被设计为"附加物"，或是作为更大的地产的附属品矗立在这些项目的近处。因此，这些别墅没有超过两个卧室的，并且室内空间都尽可能缩小了。除了纯粹的功能性考虑外，设计师明确接收到这些项目要迎合景观的要求。该项目有一个乌托邦式的入口。所有的项目都是基于简单几何形的变形，每个变形都是与室内全景相协调设计出来的，框架自然并构造出整个景观。这一特点着重强调了项目的图解——自然和大厦之间的分歧。这个别墅，从 2013 年 1 月开始设计到 2014 年 4 月，现在展现在这里的有：两幅草图、一幅操作图解、地板设计、竖向设计、剖面设计和装饰设计。当被赋予了它们的概念本质，这些项目覆盖了案例学习和娱乐之间的界限。

"Villae Minimae" (Small Villas) is a collection of five isolated single-family villas that were developed and studied at a 1/200 scale. In all five cases, the projects were designed as 'appendices,' or as additions detached from larger properties but located in the immediate vicinity of those project sites. Thus, the villas never contain more than two bedrooms, and interior spaces are minimalistic. Beyond to pure functionality, designers were explicitly requested to think of these projects as artifacts, or as machines for observing the landscape. The projects have in common a provocative and utopian approach. All projects are based on the distortion of a simple geometric figure; each distortion is a device designed to correspond to a panoramic view from the interior space, framing nature and allowing for contemplation of the landscape. This characteristic underlines the iconography of the projects: the dichotomy between nature and edifice. The villas, designed between January 2013 and April 2014, are presented here with a selection of drawings: two sketches, a diagram of the operational distortion, floor plans, elevations, sections, and renders. Given their conceptual nature, these projects occupy the border between case study and divertissement.

半山建筑
Semi-Mountain Architecture

项目地点：台湾省南投市 /Location : Nantou, Taiwan
项目面积：189.376 5 平方米 /Area : 189.376 5 m²
公司名称：杨焕生设计事业有限公司 /Organization Name : Yhs Architecture& Interior Design Co.,Ltd.
设 计 师：杨焕生 /Designer : Jacksam Yang

IAI 杰出设计大奖

杨焕生 Jacksam Yang
中国台湾 Taiwan, China

东海大学建筑硕士；
杨焕生建筑室内设计事务所主持人；
2015 年 现任亚洲大学室内设计学系讲师；
2013 年 美国《INTERIOR DESIGN》中文版杂志封面人物；
2005 年 成立杨焕生建筑室内设计事务所。

Master of Architecture Tunghai University ;
Design of Architecture Tunghai University;
Design Director Yhs Architecture&Interior Design Co,Ltd.;
2015   Lecturer of School of Design ,Asia University;
2013   The cover people of INTERRIOR DESIGN Magazune (America);
2005   Created Yhs Architecture &Interior Design Co.Ltd..

这栋建筑位于八卦山台地，视野辽阔、可以远眺中央山脉群山，也可俯瞰猫罗溪谷，宁静、优雅的文化与风土人情，随台湾现代化交通系统通讯网的便捷，在这半山与都市接轨十分方便。
委托建筑设计时，这基地附近均是大片低矮纵横排列整齐的茶园及八卦山脉特有的大片种植菠萝的菠萝园，这井然有序的绿意及远眺的高山是半山基地特有的景致，所以业主希望建筑在完成时能在室内也能欣赏这份景致。
建筑的表皮采用白色涂料，是这片绿色山林最好的配色点缀。
在设计上建筑物水平向长 30 米，是以中心二层建筑量体为主，利用框架式建筑量体设计，彼此镶嵌、重叠、错离及融合构成方式组成，由主要空间水平向延伸至右边面积为 16 米 ×3.5 米长的户外雨棚拉出建筑的前方空间，与左侧 12 米长钢结构车库顶棚形成一水平长向白色建筑量体。

The building is located in Bagua mountain mesas, vast horizons, overlooking the central mountain range , also overlooking the Maolou valley , the quiet elegance of culture and customs, along with Taiwan's modern transport system and convenient communication network, made this semi-mountain connect to the urban very convenient. When the owner commissions to design the new house, nearby this building is a large low tea plantation, hope when construction was completed, the owner could also enjoy this view inside the house.
In this design of the building, the length is 30 meters and the center is the main two-story building, the use of overlapping, composition and fusion composed by extending the right level of secondary space 16 meters times 3.5 meters outdoor rain shelter and the left side is 12 m long steel garage ceiling to form a horizontal long white building massing.

Building Space 建筑空间

华润石梅湾国际游艇会会所
CR Shimei Bay International Marina Club

项目地点：海南省万宁市石梅湾旅游度假区 /Location : Shimei Bay, Liji County, Wanning, Hainan
项目面积：5 000 平方米 /Area : 5 000 m²
公司名称：广州杨家声设计顾问有限公司 /Organization Name : Ben Yeung & Associates Ltd.
设 计 师：杨家声 /Designer : Ben Yeung

IAI 年度最佳设计机构

ben yeung & associates ltd.

杨家声建筑师事务所
Ben Yeung & Associates Ltd.
中国香港 Hong Kong, China

杨家声建筑师事务所成立超过 12 年，在香港、广州、上海均设立公司，现有专业人员 150 余人，已为许多香港及国内开发商提供服务，包括建筑、室内、景观园林设计、灯光设计、导向标识设计、品牌、软装及项目管理等。
态度改变生活是杨家声建筑师事务所的企业理念。

Ben Yeung & Associates Ltd. has been established for over 12 years. The firm has more than 150 staffs and 3 offices in Hong Kong, Guangzhou and Shanghai. "Attitude changes life" is the philosophy. Encompassing Architectural、Interior、Landscape、Lighting Design、Signage、Decoration、Master Planning and Project Management, the firm is well-known for its expertise in handling different types of projects.

华润石梅湾国际游艇会会所坐立于海南岛石梅湾区中心，整体设计体现海洋的自然魅力。会所形态仿若一双巨大的银鲸腾空出海，屋顶采用线条流畅的双曲面造型，与码头的海浪造型微妙呼应，完美、和谐地融入整个海湾环境。
鲸背的屋顶曲线设计经过多次的推敲和考量，每块量身铸造的顶部铝单板都通过精密三维空间技术准确分析，在现场三维定位后进行拼装，建筑难度与广州大剧院近似，务求从石梅湾区的每个角度都能得到会所的完美视觉体验。

China Resources Shimei Bay Yacht Club is situated in central Shimei Bay in Hainan. The overall design reflects naturally marine beauty. The club looks like a pair of silvery whales leaping out of the sea. Its double curved surfaces of the roofs create a subtle interaction with the ocean-wave- shaped docks. Consequently, the club is in perfect harmony of the bay environment.
Aiming to get a perfect visual experience from every angle of the bay, designers carefully consider the double curved surfaces of the roofs. Every customized aluminum panel is assembled on site after accurate analysis by three-dimensional space technology and localization. This step is as difficult as the one of Guangzhou Opera House.

会所立面利用弧形玻璃幕墙精心设计，强化流线轮廓，简洁、灵动、纯净。顶部利用仿生鲸背气孔设计概念，设置自然采光天窗，让光线充满整个内部空间。同时，巨型落地玻璃的幕墙设计让室内的每处主要功能区域都能体验到全开放式的海洋景观。双鲸间独创户外中轴高架平台，更将视野延伸至与之相辉映的外海入口双灯塔处，130°的广阔视野范围为使用者提供绝佳的全港湾视觉盛宴。

Made of accurate glass with sleek lines, the club facade is simple, alive and pure. There is a blowhole as a natural lighting provider on the top of each roof to fill the entire interior space with sunlight. The floor glass windows let you enjoy the amazing ocean views from every prime functional area. Joining the two buildings is an elevated platform which provides a stunning panoramic view of the ocean. The platform is creatively placed as the centre axis, leading to the entrance of the ocean, where the two lighthouse stand. You'll enjoy this 130° vision field of the sea as a visual treat.

UNASUR 秘书总部
UNASUR General Secretary Headquarters

项目地点：厄瓜多尔 /Location : The Republic Of Ecuadon
项目面积：20 000 平方米 /Area : 20 000 m²
公司名称：Diego Guayasamin 建筑事务所 /Organization Name : Diego Guayasamin Arouitectos
设 计 师：迭戈·瓜亚塞明 /Designer : Diego Guayasamin

IAI 最佳设计大奖

迭戈·瓜亚塞明
Diego Guayasamin
厄瓜多尔 The Republic of Ecuador

设计项目集中在南美厄瓜多尔的基多，包括独特性与永恒性并存的住宅、商业、企业、体育和室内建筑等。

Design projects take place in quito-ecuador(South America),where residential, comercial,coperate,sports and interior buildingts are projected,with the hignest scuptural content that aim to have their own identity with timeless attributes.

这个项目的设想是一个没有栅栏并向公共开放，能够向小区传输立体动态感觉的，连续性开放的空间环境。同样，它不仅向后制造一个通向广场的大门，同时也是出于对环境的尊重及贡献。南美国家联盟对此项目在其限制范围内分析了它的提案，以及它对社区、图书馆、剧院和休闲区开放的空间有着社会包容性的提议。

在形式上，这个新的大楼看上去像一个上升的 U 形被三个虚拟轴所限定：南北轴，La Mitad del Mundo 城市的投射和这些轴在 47°上的结合，来自北部最大偏差（23.5°）和南部最大偏差（23.5°）影响。对比之下，体积巨大，并且暗喻了自由。该建筑设计以一个雕塑的意义来展现不同的形象，这是它最令人惊喜的一个特点。

在室内设计上采用了流动空间设计,这都得益于建筑本身的外形和材料的使用。中性色的调色板和玻璃的使用增加了社会和政治的透明感的输送；创造了开放和多产的空间。

大水台的基础强调并戏剧化了悬浮在空中的整体的映射效果。

The site of the proposal has been conceived as a continuous environment without barriers and open to the publictransmitting dimensional dynamic space (proposal) to the mentioned Complex. Likewise, it is setback generating alarge entrance plaza to the project not only, but as a gesture of respect and contribution to the immediate environment.

The project for UNASUR has not only analyzed the proposal within its limits; It is a socially inclusive proposal that generates open spaces for the community, a library, theater and recreational areas.

Formally the new building appears as a volume like an ascending U delimited by three virtual axes: The North-South axis, the projection of the proposal with La Mitad del Mundo City and the combination of these axes in 47 degrees, resulting from the sum or displacement of maximum declination north (23.5 degrees) and maximum declination south (23.5 degrees) of the sun in relation to the equator.

The volume is a massive piece by contrast rises and is projected as metaphore of freedom. The building has been designed as a sculptural connotation to surprise and show a different image depending on where it is appreciated.

The interior design proposes fluid areas, thanks to its configuration and use of materials. A palette of neutral colors protocol is used. The use of glass adds to the idea of conveying a message of social and political transparency; peaceful, open and productive spaces without major distractions.

The large body of water like base accentuates and dramatizes the reflection effect of a monolithic body suspended in the air.

北京不可能空间
Impossible Space Beijing

项目地点：北京798国际艺术区B区706北街B08-3号 /Location : No.B08-3, North Street, District B 706, 798 International Art Area, Beijing
项目面积：276平方米 /Area : 276 m²
公司名称：墨匠国际建筑设计顾问（北京）有限公司 /Organization Name : Inkmason International Architectural Design(Beijing)Co., Ltd

IAI 设计优胜奖

墨匠国际建筑设计顾问（北京）有限公司
Inkmason International Architectural Design,Beijing,Co., Ltd

墨匠国际作为一家充满活力的多元化设计公司，其专业团队汇集了一群富有热情，追求卓越的建筑师、室内设计师和项目经理。
他们秉承"以墨为思、以匠为行"的理念，致力于为客户提供专业、完善的设计和工程服务。
墨匠与国际化企业客户紧密协作，以客户企业文化和愿景为方向，以使用方需求为基准，为其提供量身定制的设计方案和工程服务。

Inkmason is a dynamic multicultural team of architects, interior designers and project managers. Together they unifiedwith passion and pursuits for excellence. they collaborate with their clients to ascertain their needs and perspectives. By thoughtfully understand their corporate culture and vision.
They provide tailored solutions. That leads to a rewarding outcome of design solution with team dedication and reflects their values, laying foundations for long-standing partnership with reputable international clients.

林健强 Lin Jianqiang
中国 China

资深建筑设计师，从事建筑设计、酒店室内设计、运动设施项目管理，以及工业园区的总体规划。

Lin Jianjiang, senior architect, engaged in architectural design,hotel interior design,sports facilities project management and the general layout of industrial park.

项目位于北京798艺术区，原本仅是一座2层旧砖结构的军工厂房遗址，但经过整体的改建，工厂变成了一座充满了青春气息的艺术画廊。
事实上，改造一间废弃的厂房，并把它赋予时代的气息这就如同"Impossible Project"企业发展的故事一样——拯救即将关闭的工厂，给产品赋予"即时和方便"的功效并与艺术设计相结合。
Inkmason的设计团队想要通过保留建筑的原始砖结构来强调建筑的历史意义，最终通过将新旧材料的碰撞与结合的方式来打造。建筑外墙上，各种新旧材料以层次并列的方式加上不同透明度的材料，使人们不仅可以看到原本的砖结构，还可以欣赏到新旧材料的和谐结合，使建筑物焕然一新。

In Beijing 798 Art District, Inkmason was able to give a new life to this abandoned 2 stories old brick structure and transform it into a modern art gallery .The fact that renovating an abandoned brick house and bring it up to date with a much more elevated purpose is so much like the" Impossible Project "story – revitalizing a closing factory and transforming something "instant and convenient" into something artistic.
Inkmason's architect wanted to emphasize the tactile feeling from the original brick structure which reminds people of the historical significance of the buildings in the district. In the end the solution is interlocking and intertwining the new and old structures creating interlocking blocks with new and old materials coming in and out of each other. This is being done by using large amount of steel framed glass panels at the façade which allows people to seen the brick structure inside yet able to appreciate the glass envelope.

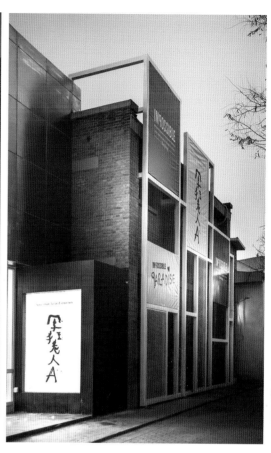

Building Space 建筑空间

时代美术馆
Times Museum

项目地点：广东省广州市 /Location : Guangzhou, Guangdong
项目面积：5 773 平方米 /Area : 5 773 m²
公司名称：CL3 建筑师有限公司 /Organization Name : CL3 Architects Limited
设 计 师：林伟而 /Designer : William Lim

IAI 设计优胜奖

林伟而　William Lim
中国香港　Hong Kong, China

美国麻省注册建筑师；
美国建筑师注册管理局注册建筑师；
中华人民共和国一级注册建筑师资格；
美国康乃尔大学建筑系学士、硕士；
香港思联建筑设计有限公司董事、总经理；
在美国波士顿工作了五年后，于 1987 年回港，
在一间香港地产发展商工作了六年；
除作为一名建筑师外，他还活跃于艺术领域，主
要是公共艺术、摄影和绘画方向。

Registered architect of Massachusetts in the United States;
Registered architect of American Architect Registration Administration;
The 1st class registered architect certificate of the People's Republic of China;
A master's degree of Cornell university bachelor of architecture;
Managing director of Hong Kong, united architectural design co., LTD;
After working in Boston, USA for five years later, back to Hong Kong in 1987, working in a property developers working for six years in Hong Kong;
Except as an architect, he was also active in art, mainly public art photography and painting.

时代美术馆是利用 1980 年代建于中国南方广州市提供给工厂工人的四栋房子进行改造而成的。原建筑一直空置且无人关注，此项目将它的首两层改为一个当代艺术博物馆，并有相邻的书店和咖啡馆，四层则变为设计师的工作室。以可持续的方式设计该项目，最大限度地利用自然通风，创建一个巨型顶棚遮盖全部（四栋）楼宇，用开放式的玻璃和金属桥梁和新安装的电梯将它们连接起来。

Times Museum is an adaptive reuse project occupying 4 documentary blocks used to house factory workers in 1980's Guangzhou, Southern China. Having been unoccupied and gone to neglect, the project is to convert the buildings into a museum for Contemporary Art on the first 2 levels, with an adjoining bookshop and cafe, and 4 levels above that into studio workshops for designers. Using a sustainable approach to the project, the use of natural ventilation is maximized be creating a large canopy on top to shelter the 4 buildings, and open glass and metal bridges to connect them to newly installed elevators.

GROUND FLOOR PLAN

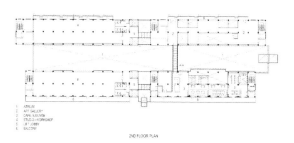
2ND FLOOR PLAN

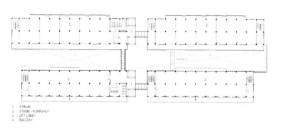
3 - 6TH FLOOR PLAN

现有的露天通道被金属网封闭以保留自然的空气流通。这可以令整个公共区域无需安装空调系统。工作间保留原有可开启的窗户和通往阳台的门。尽量减少使用新的墙体饰面，内部空间采用金属、玻璃桥和金属网罩，形成自然原始的对比。三组大型的竹子群，增添了大自然的盎然绿意，而戏剧性的扩散照明为内部空间营造出一份宁静街区之感，街区末端为体育场阶梯式休息区，可同时用于艺术展览。

Existing open walkways are enclosed with metal mesh to retain natural air flow. As a result, no air conditioning system is installed for the entire public area. The studio workshops retain existing openable windows and doors onto balconies. Keeping use of new wall finishes to a minimum, the interior spaces has a raw look contrasting the metal and glass bridge connections and the mesh enclosures. 3 large clusters of bamboo planting add a sense of green nature, and dramatic diffused lighting gives an Asian tranquility to the long internal avenue, terminating in a stepped stadium seating area which can also be used for art display.

竹石山房——安吉山川乡游客中心设计
Mountain House——Tourist Center of Shanchuan, Anji

项目地点：浙江省湖州市安吉县山川乡霞大线、双后线 /Location : Shuanghou Line and Shuagshui Line, Anji County, HuZhou, Zhejiang
项目面积：2 360平方米 /Area : 2 360 m²
公司名称：杭州聚石建筑设计事务所有限公司 /Organization Name : Hangzhou Jushi Arcvhitects Co., Ltd.
设 计 师：康胤，徐亚东，张驰，王斌，于劭扬 /Designers : Kang Yin, Xu Yadong, Zhang Chi, Wang Bin, Yu Shaoyang

IAI 设计优胜奖

浙江省安吉县山川乡的竹石山房是一座隐于自然环境之中的游客接待中心，我们提取了山川乡当地的"高山""竹林""梯田""石屋"四大标志性元素作为设计来源，结合具体场地，将四大自然元素以拟山形式、地形渗透、地域材料使用和屋顶造林的手法运用在建筑设计中，最终完成了这座与自然相融，传递地域特征，并使人可游乐于其间的竹石山房。

The mountain house is a tourist center located at the entrance of Shanchuan country, Anji county, Zhejiang province.Designers abstract the most distinguishing elements of local feature —mountains, bamboo forests, terraced fields and traditional stone houses—as the soil of the design. Then designers put them into the detailed site context, use the methods of mountain-like form, topography permeation, local material using and roof garden creation in their design, and finally create this natural, regional and interesting mountain house.

杭州聚石建筑设计事务所有限公司
Hangzhou Jushi Architects Co., Ltd.
中国 China

杭州聚石建筑设计事务所有限公司位于杭州西子湖畔，事务所前身于2001年组建成立，于2012年更用现名。
事务所现有人员30余人，公司业务范围涵盖建筑工程设计、规划设计、景观设计。目前建成项目均在当地获得良好的社会反响。
公司以"卓越时尚的设计理念，体系完美的图纸质量，至诚合作的现场服务"为服务宗旨，创作更多的设计精品。

Hangzhou Jushi Architects Co., Ltd.is located in Hangzhou West Lake,the firm's predecessor was founded in 2001 and changed the existing name in 2012.
The existing business staff has more than 30 people,and the scope covered architectural engineering design ,planning and design ,landscape design.The projects all receive good reviews.
The company in the service cooperation of "outstanding fashion design concept ,perfect system drawing quality ,n-site service of sincere cooperation",to create more good designs.

横琴杨昶博物馆
Hengqin Yangchang Museum

项目地点：广东省珠海市横琴开发区小横琴山 /Location : Hengqin Development Zone, Zhuhai, Guangdong
项目面积：80 000 平方米 /Area : 80 000 m²
公司名称：杨侸环境艺术设计有限公司 /Organization Name : Zhuhai Yang Er Environment Art Design Co. Ltd.
设 计 师：杨侸 /Designer : Yang Er

IAI 设计优胜奖

杨侸 Yang Er
中国 China

杨侸环境艺术设计有限公司创办人、首席设计师；
深圳室内设计师协会常务理事；
从事设计工作 20 年；
2013 年中国国际大学生空间设计大奖 "ID+G" 金创意奖特邀专业导师；
2005 年创立杨侸环境艺术设计有限公司；
2003 年进入香港高文安设计公司，期间任职手绘主笔、设计总监；
1995 年先后在北京、珠海创立个人设计公司。

Chief Designer and Founder of Yang Er Environment Art Design Co. Ltd.;
Executive Member of Shenzhen Association of Interior Designers;
2013 Professional Instructors of "ID+G";
2005 Created Yang Er Environment Art Design Co. Ltd.;
2003 Design Director of Hong Kong Kenneth Ko Design Co.,Ltd.;
1995 Created personal design company in Beijing and Zhuhai successively.

项目以圆形为主体，被莲花池包围，孤出宁静，显现出世外之境的东方禅意。其心理动线根据中国的人文哲学，运用迂回隐喻的路径，创造又一山和世外之源的想象空间的体验感。先通过商业综合体才能进入，从博物馆出来必须经过商业综合体，以及文化创意广场和有文化创意形象的建筑，使整个环境有生活、有活力、有生命之养。引入山泉小溪之水至弧形围墙的顶部蓄水，在内院形成20 米高的落水瀑布让人意外和震撼，在美观的同时又起到对室内降温的作用。弧型围墙形态的博物馆摄入山体，环抱交融，也把自然的山石借入到室内，使人文环境与自然合而为一。水泥三棱柱结合，罗列咬合结构，组成容器壁。

The round body, is surrounded by the lotus pond, solitary and quiet, as the Oriental Zen of the wonderland.According to the China humanistic philosophy, using the roundabout route, it creates an experience of the Imagination space with another mountain and wonderland. Firstly,through the commercial complex into the museum, must go through a commercial complex from the museum, and the cultural and creative Plaza and cultural and creative architectural image, so that the whole environment has a life, has the vigor, is a life support.Introduction of mountain spring stream to arc wall at the top of the water storage, water, courtyard form 20 meters high waterfall drowning surprise and shock. At the same time for the beautiful and of indoor cooling effect. Museum in the form of arc round intake mountain, surrounded by blending, also the nature of the rocks to borrow to indoor, the humanities environment and the natural one.Listed cement system of three prism, interlocking structure, composed of vessel wall.

 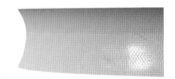

印象武隆小镇规划设计
The Impression of Wu Long Primitive Town

项目地点：重庆市武隆县 / Location : Wulong District, Chongqing
项目面积：104 666.6 平方米 / Area : 104 666.6 m²
公司名称：西南大学美术学院 & 建筑艺术研究所 / Team or Organization Name : Academy of Fine Arts & Institute of Architecture, Southwest University
设 计 师：张汉平 / Designer : Zhang Hanping

IAI 设计优胜奖

张汉平 Zhang Hanping
中国 China

2008 年 香港华发装修工程 (重庆) 有限公司主任设计师暨艺术总监；
2007 起至今 任职西南大学建筑艺术研究所，副所长；西南大学美术学院艺术设计系教师；
2006 年 《瞭望》杂志 / 中华脊梁特辑《建筑艺术家张汉平》；
2004 年 著有《设计与表达》北京计划出版社，并出版；
2004 年起 任海南金厦装饰工程公司 - 主任设计师；
1996 年 于重庆开办设计事务所；
1991 年 毕业于西南大学建筑艺术研究所获建筑艺术硕士学位；在海南开办设计事务所。

In 2008 Designer and art director in Hong Kong huafa decoration engineering (chongqing) co., LTD;
Since 2007 Deputy director of the institute at the University of Southwest Architectural Art;Since 2007, deputy director of the institute at the University of Southwest Architectural Art,teacher in Southwest university academy of fine arts department;
In 2006 The Outlook Magazine/the BacTbone Special Building artist han-ping zhang.
In 2004 Write and Publish Design and express in Beijing plan press;
In 2004 Chief designer in Hainan Jinxia Decoration Engineering Company;
In 1996 Set up design office in Chongqing;
In 1991 A master's degree in art and architecture,Graduate from Southwest University Institute of Architectural Art,Set up design office in Hainan.

1. 《印象·武隆》是张艺谋执导的大型实景山水歌舞剧的表演艺术延伸舞台；
2. 情景 / 体验式旅游风情镇；
3. 乌江．峡江．武陵文明原生态山地场镇；隐性人类文化遗产保护；开放式活态博物馆；
4. 川东武陵山区最美乡村古镇单 / 群体土木古建筑集萃迁建；
5. 武隆 - 仙女山旅游新的节点，景点，新景区开发游客中转集散地；
6. 环境友好型 - 绿色生态宜居小镇（核桃镇 + 印象武隆小镇）；
7. 美丽中国．美丽乡村．新城镇化改革试点。

1.Theme and goal companion piece of musical drama "Impression of Wu Long" directing by Zhang Yi Mou.
2.Scene and experiential tourist town
3.The original ecological mountain town of Wu Jiang river civilization,The recessive material culture heritage:open and active state museum.
4.Concentrate most beautiful village and town of east part of Si Chuan province.
5.The new scenic spots of Wulong Fairy mountain tourism,Tourists transit hub
6.Green ecological livable town.Trial reform of the new urbanization:
7.Beautiful China,Beautiful Country ,trial reform of new town.

Industrial Product
工业产品

Read & Life	284
"Sit on it"	286
"Beat the Chair"	287
"Repairer"	288
"Piggy" Bank	289
Ring Ceremony	290
Shine Go Portable Office Desk	292
"The Star of Changsha" Sculpture	294

284	喜阅人生
286	"坐下去"
287	"打坐"
288	"修理工"
289	"福猪"储蓄罐
290	环礼
292	炫 Go 办公桌
294	"长沙之星"雕塑

喜阅人生
Read & Life

公司名称：厦门凤飞服饰设计有限公司 /Organization Name : Xiamen FENGFEI Fashion Design Co., Ltd.
设 计 师：纪锋思 /Designer : Ji Fengsi

IAI 最佳设计大奖

纪锋思　Ji　Fengsi
中国　China

2010 年至今　厦门凤飞服饰设计有限公司
　　　　　　北京曾凤飞时装设计有限公司
　　　　　　曾凤飞助理兼首席设计师

Since 2010,Xiamen FENGFEI Fashion Design Co., Ltd.;Beijing Ceng Fengfei Fashion Design co., LTD;Assistant and Chief Designer of Ceng Fengfei .

本系列"喜阅人生"作品，以愉悦怡然的心境总结阐释了喜阅人生的处世态度。"喜阅人生"，时刻心怀欣喜，善于发现生活中的点滴美好，并乐于分享这份美好，传递了积极乐观的豁达人生观。本系列以惯有的手法对中国传统文化元素进行提炼，于自然风景、日常生活场景取材，糅合现代时尚理念，融入设计师所要传递的喜阅人生百态的随性、豁达态度，探寻中国传统文化根源所蕴藏之人生智慧。设计风格上延续品牌一贯的现代中式风格，在款式细节、结构造型、图形图案、面料创新设计等方面都力求回归服装本身，提升服装穿着舒适度和服装功能性，重在对服装结构、面料创新的研究。将传统中国元素主旋律和谐而富有节奏的特点融入于本季服装系列设计中，深化品牌内涵。

This series of "Read & Life" works, the joy of happy mood summary explains the attitude towards life like reading. Read & Life, with joy at the moment, found that the goodness of life bit by bit, and willing to share this happiness, convey positive and optimistic open-minded outlook on life. This series with the usual methods of Chinese traditional culture elements are refined, from the natural scenery, daily life scenes, blend into modern fashion concept, designers have to pass the vicissitudes of life with joy to open-minded attitude, explore the hidden China traditional cultural roots of the wisdom of life. The design style of modern Chinese style has always been the continuation of the brand positioning, in terms of style, structure, pattern and other details, fabric innovation design and strive to return all the clothing itself, enhance the clothing comfort and taking functional, to study the structure of clothing, fabric innovation on. The traditional Chinese main melody and rhythm harmonious elements into the season clothing design, deepen the connotation of the brand.

Industrial Product 工业产品

"坐下去"
"Sit on It"

项目尺寸：50厘米×50厘米×65厘米 / Size : 50cm×50cm×65cm
设 计 师：贾蓓蓓 / Designer : Juliet Jia

IAI 最佳设计大奖

贾蓓蓓 Juliet Jia
中国 China

2014年至今　北京城市设计学院艺术系家居产品设计教师
2014年　北京工业大学耿丹学院艺术系家居产品设计教师
2011年　作品"打坐"入围第五届"为坐而设计"，被伦敦艺术大学设计师Ben Hughes收藏。并分别刊登于《AD china》（2012）和《ELLEDECO家居廊》（2013）

Since 2014　Household product design teacher of art department of Beijing institute of urban design;
2014—2014　Household product design teacher of art department of Institute of Beijing university of technology GengDan;
In 2011"Sit" work shortlisted the fifth award "design for sit",collected by University of the arts London designer Ben Hughes,and respectively in the issue of AD China (2012) and ELLEDECO household gallery (2013).

通过利用敲打生活中的废旧金属制品再设计成为一把椅子的过程，来探索人们对于使用生活中物品的心理和反应，人的行为与材料之间的关系和材料本身的故事。

贾蓓蓓对人的行为与材料之间的关系很感兴趣。这个作品最初的灵感来源是自然中的风和乐手敲打的架子鼓，她想设计一把椅子可以发出声音，在人们体验时可以敲打。初步设想是收集一大堆废旧金属制品对其进行随意敲打，直至敲打出舒适的可以坐的坐具。后来在材质上选用了人们生活中常见的不锈钢制品——锅碗瓢盆，她的设计开始更多地与人相关。很重要的一部分设计是在马路上来制作，同时设置一架照相机记录人们对于她的行为的反应。每一部分材料都由她亲自收集，在路边敲打造型，与生活中的普通人交谈，从中获取人们对于材料和物品的看法，从而进一步感受材料本身的含义。整个过程中"完成"了人的行为与材料之间的对话。

Juliet Jia wants to through a process which by beating waste metal projects in daily life to explore how people change their psychology and reflection when they use things. What is the relationship between people's behavior and material, and the narrative behind the material.
I am interested in the relationship between people's behavior and material. The first inspiration is come from wind in nature and drum. She wants to design a chair can make voice the make people beat it. The material She chooses steel metal projects from daily life-pans and pots. The most important design is she make it on the public road. She records people's reflection by a camera during the process. She beats every part of the material, talk with people, recording their thinking about my work, then she thinks about material's meaning inside.

"打坐"
"Beat the Chair"

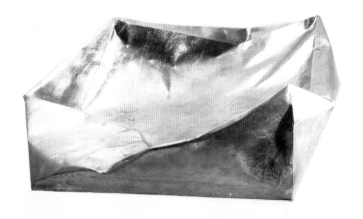

项目尺寸：100厘米×90厘米×10厘米 /Size：100cm×90cm×10cm
设 计 师：贾蓓蓓 /Designer：Juliet Jia

IAI 设计优胜奖

作品的灵感来源于在被人们的手或脚捏过踩过，或在路边被汽车碾压过的易拉罐。这些易拉罐常常被人丢弃，但事实上它们是具有美学价值的。人们常常在事物上留下痕迹，它们背后的故事往往反映着人们的心情、行为和习惯。

贾蓓蓓的家具邀请人们通过在薄金属上随意的试坐来创造属于他们自己的椅子，记录生活中瞬间的痕迹，再用传统的金属翻铸的手法，记录人们留下的痕迹，将瞬间的痕迹凝聚为坚不可推的金属。同时将长达一个月的翻铸过程与一次性的痕迹形成对比。为突出坐痕，在设计的座椅底座部分选用透明亚克力材质。同时，利用摄像来纪录人们在薄金属上留下痕迹的过程，解释并且定义不同个体的痕迹留在材料上的意义。此次展品为此设计的一部分，在伦敦艺术大学的创作中共有四件。

The inspiration is waste cans squeezed by people's hands, feet, and by cars on the road. It is actually beautiful however it is always ignored. People always leave a trace on things. Its underlying narrative reflects people's mood, actions and habits.

Juliet Jia 's furniture invites individuals to create their own chair by sitting freely on thin metal, retaining the momentary trace of life. The way she keeps the trace is using foundry with aluminium. The long making process can be a contrast with just disposable trace. She uses acrylic is because she wants to just make the sit pattern stand out. She tries to record the process people leave trace on the thin metal by video, to explain and identify the meaning of different individual's trace on the material.

"修理工"
"Repairer"

项目尺寸：110厘米×90厘米×110厘米 /Size : 110cm×90cm×110cm
设 计 师：贾蓓蓓 /Designer : Juliet Jia

IAI 设计优胜奖

修理工这一人群的社会地位很低，他们的生活质量往往也很差。但在我看来他们的手工艺技能是高超的。这种工业社会的环境影响了她很多，所以她决定通过这件作品使人们注意这个人群并反思修理工的社会地位。

通过深入到修理工的生活中，收集他们的故事，每日从报废汽车厂收集零配件并进行清洗分类，再利用修理工的技能将其改装拼接在一起完成设计。除主体结构部分和靠背部分焊接，其他所有零配件可拆解，座椅可以360°旋转，后靠背可以折叠合拢，三个侧面设有抽屉。座椅可以与人形成互动空间，就像观众自己为一个修理工一样。

The social status of repairers in China is very low and their life quality is usually very bad. However, Juliet Jia thinks their handmake skill is amazing. So she decides to do this work to make people reflect and rethink repairer's social status and notice this kind of people.

Juliet Jia did research in many repair factories and talked with repairers, collected their stories. At the same time she collected waste car part materials from waste factories everyday, and clean them when she came back home. I found a local repairer help me to make the chair. she used repair's handmake skill to do the work. Except the main structure was weld, other parts can be took off. Every part of the chair comes from car. And the chair can roll in 360 angels, the back of the chair can fold, there are three drawers on the side of the chair. When people sit the chair, they can play with the small material just like they are repairers.

"福猪"储蓄罐
"Piggy"Bank

项目尺寸：160厘米×160厘米×160厘米 /Size : 160cm×160cm×160cm
设 计 师：黄信尧 /Designer : Xinyao Huang

IAI 设计优胜奖

黄信尧　Xinyao Huang
中国　China

毕业于江西景德镇陶瓷学院雕塑专业；
2008年至今　任职于标致雪铁龙亚洲研发中心创意雕塑师；琅沐创意年代微信公众平台，创始人；主导负责标致雪铁龙亚洲研发中心多款车型改款；
2006—2008年　任职于上海浩汉工业产品设计有限公司；交通工具设计部任油泥造型师。

Graduated from jingdezhen ceramic institute sculpture professional;
Since 2008,work as a creative sculpture in Psa Peugeot Citroen r&d center in Asia;
Founder of Langmu WeChat Creative public platform,Main responsible for psa Peugeot Citroen Asia r&d center model modifying;
2006-2008　work as a Clay modelling designer in transport tool department in Shanghai HaoHan industrial product design co., LTD.

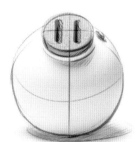

环礼
Ring Ceremony

项目尺寸：2400毫米×600毫米×400毫米 /Size : 2400mm×600mm×400mm
设 计 师：陈丽君 /Designer : Chen Lijun

IAI 设计优胜奖

陈丽君 Chen Lijun
中国 China

中央美术学院城市设计学院家居产品专业09级本科学生；
现中央美术学院城市设计学院家居产品专业研究生在读；
2015年1月至7月 在丹麦柯灵设计学院进行交换生学习。

获得由中央美术学院主办的 2012 年 "天鹤奖" 中国国际青年设计师大赛银奖；
作品《环礼》收入《创意150——中央美术学院城市设计学院家居产品专业毕业作品集》。

The 09 level of undergraduate student,Home products professional of the school of urban design,Central Academy of Fine Arts.Professional graduate students reading in home product professional of the school of urban design,central academy of fine arts.
In January 2015-July kolling design institute for the exchange students study in Denmark.

Geting sponsored by the central academy of fine arts crane award for 2012 days China international youth designers contest silver ring ceremony .
Works Ring Ceremony has been collected into The Creative 150——the graduation portfolio of household products design in urban design college of The Central Academy of Fine Arts.

现代人对于礼制的认识是在逐渐淡化的，人们对于长幼尊卑、主客之别的意识也在各种文化的冲击下变得界限模糊。《环礼》系列作品是以龙山黑陶为物质的表现形态，借助其独特的关于"礼"的意象语言，融入传统文化的感召力，重新将黑陶定义，让现代人再一次认识黑陶、认识传统文化。在使用器物的同时，唤起礼制记忆的复苏，在一颦一笑间，感受到礼制的影响力，使得行为本身，就像是在诉说一个发人深省的故事，或一个浅显易懂的道理，而黑陶，也不再单单是一件器物，它架起了现代与古代的桥梁，扮演了一个沟通者的角色，让礼制的精神得以传承和发扬，引起人们进行更深入的思考。

Nowadays, it is a common phenomena that people's attitude towards etiquette drops down gradually. Under the impact of various culture collisions, people's awareness of etiquette becomes more and more blurred. Series of work--"Ring Ceremony", based on Long Mountain's pottery as the material and unique imagery language of etiquette, refining the modern meaning of etiquette. What's the most important is the inner meaning through this product.Chen Lijun sincerely hopes that her design could wake up modern people's attention to our Chinese traditional culture, apart from that, she believes that "Ring Ceremony" would become a bridge which could link the ancient and the present, a inheritor whom could inherit and develop the traditional courtesies, and a mirror that could enlighten and deepen peaple's inner minds.

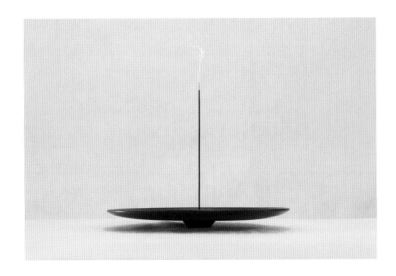

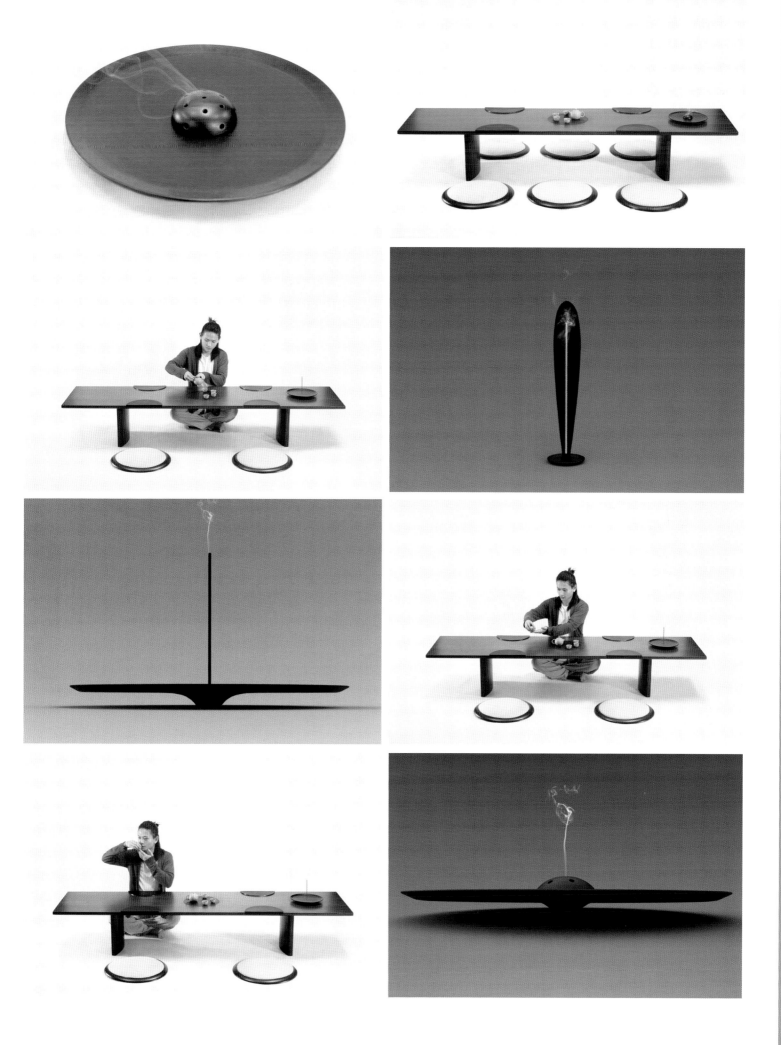

炫 GO 办公桌
Shine Go Portable Office Desk

项目尺寸：1800毫米×700毫米×750毫米 /Size：1800mm×700mm×750mm
公司名称：珠海天王空间设计有限公司 /Organization Name：Zhuhai Tian Wang Design Co., Ltd.
设 计 师：张成荣 /Designer：Zhang Chengrong

IAI 设计优胜奖

张成荣　Zhang Chengrong
中国　China

珠海天王空间设计有限公司总经理、首席设计师；
毕业于重庆师范大学首届摄影专业；
概念空间、建筑及景观首席设计师；
办公家具行业品牌整合策划资深推广人士；
曾任北京理工大学空间设计学院客座教授。

General Manager and Chief Designer in Zhuhai Tian Wang Space Design Co., Ltd.;
Graduated from Chongqing Normal University, the first professional photography speciality;
The chief designer in architecture and landscape concept space;
Brand integration marketing senior people in furniture industry;
He was a visiting professor at the Beijing university of technology institute of space design.

本案设计理念是以现有物品、现成材料，经过简单的构成方式，形成新的具有功能性的用具。本设计遵循因地制宜，就地取材的原则，选择有使用价值的现有材料，或废弃物为主要材料，在使用过程中不产生新的污染。通过此设计，设计师倡导"健康、环保、简生活"的生活理念。本案名为"炫go"。"炫"一方面源于本案设计上的创意，时尚、简约、动感，充满活力；另一方面表达了年轻一代勇于打破传统工作与生活的界限，追求健康、时尚、动感的工作和生活方式。"go"是一种不断向前、积极向上的精神，在形象上，"go"的两个圈"o"与自行车的两个滚动车轮相同。"炫go"就是让色彩炫起来，让生命动起来。

This design involves in reconstruction of an old building. The beams and extremely sophisticated pillars of the former building allow no changes or reconstructions, by taking reference from the design theme, the designer makes use of disadvantages of the building as design elements and brings new life to the brand culture of JONGTAY with modern techniques. JONGTAY takes "harmony culture" as the corporate culture, with "harmony brings everything" as the essence of JONGTAY's brand culture. Among diversified Chinese cultural elements, elements fitting best to conceptual design, including the six directions (six harmonies), svastika, lotus flower and grids are selected, together with the dominant hue acted by Chinese red, shocking visual effects full of infectivity are created. As for design, modern styles and modern materials are used mainly to deliver the connotation of the traditional "harmony" culture of China. Perfect expression of brand culture and space design is realized through the organic transition between VI corporate image and space.

"长沙之星"雕塑
The Star of Changsha (Sculpture)

项目地点：湖南省长沙市梅溪湖节庆岛西侧一角 /SLocation : Meixi Lake Holiday Island, Changsha, Hunan
项目尺寸：高18米，长19米 /Area : 18 meters tall, 19 meters long
公司名称：广东省集美设计工程有限公司 /Organization Name : Guangdong Jimei Design and Construction Co.,Ltd.
设 计 师：黄建成、何为、余鹏、方伦磊 /Designer : Huang Jiancheng, He Wei, Yu Peng, Fang Lunlei

IAI 设计优胜奖

广东省集美设计工程有限公司
Guangdong Jimei Design and Construction Co.,Ltd.
中国 China

广东省集美设计工程有限公司，隶属广州美术学院，是全国艺术高等院校最早成立的设计和施工企业（具有建筑装饰工程设计甲级、建筑装饰工程承包乙级资质）。
集美公司开创了由中国（大陆）人自己设计的全国最早的影视广告、最早的五星级酒店、最早的美术馆和最早的音乐厅等先河。进入新世纪以来，集美公司深度参与了2010年上海世博会国家馆、主题馆，广州亚运会和深圳世界大学生运动会的视觉系统设计等国家重大工程的设计工作。集美公司成立近30年来，在建筑、室内、景观、展览展示、广告、产品、服装、陶艺等领域中，拥有国家级专业学会（协会）的副会长、常务理事、理事等社会地位，已成为与中国高校系统"产学研"相结合的最成功的典范与标杆，成为中国乃至全国实力雄厚且具特色的文化创意产业基地。

Guangdong Limei Design & Construction CO.,LTD, affiliated to Guangzhou Academy of Fine Arts, is the first design and construction enterprise established by art institution of higher education throughout China. They own Grade-A Qualification Certificate for Architecture and Decoration Design, and Grade-1 Qualification Certificate for Architecture and Decoration Contracting. They made the first video advertisement, first 5-star rated hotel, first art museum, first concert hall and etc. ever done by mainland Chinese. Entering into this new century, their participated in-depth into the design part of important projects nationwide, such as the China Pavilion and Theme pavilion of 2010 Shanghai World Expo, visual system design for 2010 Guangzhou Asian Games and Universiade 2011 Shenzhen.Since nearly 30 years of their establishment, many of our designers have been holding titles like Deputy Director, Managing Director, member of Board of Directors in professional societies/associations at national-level in the areas of architecture, interior design, landscape, exhibition display, advertising, product design, fashion design, ceramics and etc. They become the most successful example and a model of knowledge Transfer in China's higher education system, and the strongest cultural creative industry base with unique characteristic in southern China and even the whole country.

创意构想为星宿（长沙星）——星云、星空、流星、星河，沙体——颗粒状、体块状、成片状，星与沙的结合——过渡衔接。
一颗星星代表一个长沙得名的含义，将这五颗星星环绕一圈形成长沙之星。形态呈散点状、云团状，概念几何图形或可以说是天体，它所代表的含义是闪耀发光。而它的状态是闪烁的、运行中的、变换的、动态的、时空的、有距离的。
设计创意来自于长沙得名的由来，长沙之名分别得名于长沙星、万里沙祠、"沙土之地"、长形的沙洲、"蛮越"语"祭礼女神的地方"。

Creative idea:the stars (changsha) ——Nebula star shooting stars，sand body —— Granular body block slice,Connection with star and sand——Conversion linking.
A star represent a meaning of changsha named,so with this five stars around it form the star of changsha.Its shape in points dystained,clouds,concept geometry or celestial body ,represents the meaning of glory and shiny.And the state of it is flashing ,running ,changing,dynamic,spatiotemporal and the sense of distance.
The design ideal is coming from the name of "Changsha",Changsha named from Changsha Star,Wan Lisha shrine,the land of sand,the long bar,the place hold a memorial ceremony for the goddess in Manyue Language,one star represent the meaning of Changsha,and these five stars form the Changsha Star.

Project Design and Others
方案设计和其他

View from Peephole	298
Beijing Liming Office Furniture Center	300
RE-DEEM: Transitional Hub for Former Prisoner	304
Air Breathing Centre	306
Lettuce House: Sustainable Lifestyle Lab	308
BANSHANYIHAO Villa Interior Design	312
Zhangjiagang Town River Reconstruction	314
Fortune World Commercial Space Decoration	316

298	窥世
300	北京黎明办公家具中心
304	赎回：前罪犯过渡中心
306	空气呼吸中心
308	生菜屋——可持续生活实验室
312	半山一号别墅室内设计
314	张家港小城河改造
316	财富天地商业空间软装

Project Design and Others 方案设计和其他

窥世
View from Peephole

项目地点：湖南省长沙市望城区龙湖湘风星城 / Location: Longhu Xiangfeng Xingcheng, Wangcheng District, Changsha, Hunan
项目面积：220 平方米 / Area: 220 m²
公司名称：鸿扬家装 / Organization Name: Hirun Company
设 计 师：贺丹 / Designer: He Dan

IAI 最佳设计大奖

贺丹　He Dan
中国　China

2011 年至今工作于鸿扬家装；
2010—2011 年初工作于长沙八九装饰设计公司；
毕业于湖南商学院环境艺术专业；
长沙市鸿扬家装设计师。

Since 2011 Work at Hirun Company;
2010-2011 Work at Bajiu Decorate Company;
Graduate from Department of Environmental Art
of Hunan University of Commerce;
Changsha Hongyang Home Decoration Designer.

本方案在原有的建筑结构上做了改造，建筑外观结合中国传统园林建筑的营造方式，融入自然、讲究虚与实、明与暗、人工与自然的相互转换，同时结合了中国传统园林的借景、对景、分景、隔景等手法重新组织空间。室内空间将中国传统的竹编形式进行改造，利用木材的韧性重新加以组合编制，造成空间多变、小中见大、虚实相间的效果，而这正是"窥"的一种体现。"窥"的解释是指从小孔、缝隙或隐蔽处偷看，窥探、窥伺、窥测、窥视。从汉字的直白意思，可以看出，"窥"是个中性偏贬性的词语。可贺丹觉得"窥"一直有贯穿于在人们的周围。窥有一种含蓄去看的意思，而含蓄本身就是东方美的一种表现。

There are some changes on the original architecture structure in this design. The way of the architecture appears also combines the way the traditional Chinese architecture built. It looks like part of the nature while paying attention to the integration of virtual and real, bright and dark, artificial and nature. Meanwhile, it also learns from view-borrowing, view-facing, view-dividing and view-separating and of the traditional Chinese architecture and re-organize the space. The inner space reforming relies on the traditional Chinese tool bamboo weaving,we take advantage of the wood's toughness to recombine it, which is easy to change and make the space appears in different ways, looking bigger though small and integration of virtual and real. This is how "peephole" shows. "Peeping" means looking though small holes, cracks or hide-ways, ad we can tell from the literate meaning that "peeping" is somehow negative. But He Dan think "peeping" is everywhere in our lives. "Peeping" is to look in an implicitly, and implication is one of the ways Oriental Beauty shows.

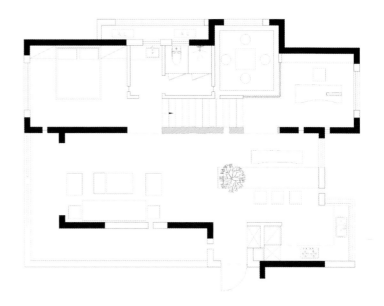

北京黎明办公家具中心
Beijing Liming Office Furniture Center

项目地点：北京市通州区马驹桥金桥科技产业基地景盛南四街18号 /Location: 18 Jingsheng Nansi Road, Tongzhou District, Beijing
项目面积：10 000 平方米 /Area: 10 000 m²
公司名称：珠海天王空间设计有限公司 /Organization Name: Zhuhai Tian Wang Design Co., Ltd.
设 计 师：张成荣 /Designer: Zhang Chengrong

IAI 最佳设计大奖

张成荣 Zhang Chengrong
中国 China

珠海天王空间设计有限公司总经理、首席设计师；
毕业于重庆师范大学首届摄影专业；
概念空间、建筑及景观首席设计师；
办公家具行业品牌整合策划资深推广人士；
曾任北京理工大学空间设计学院客座教授。

General Manager and Chief Designer in Zhuhai Tian Wang Space Design Co., Ltd.;
Graduated from Chongqing Normal University, the first professional photography speciality;
The chief designer in architecture and landscape concept space;
Brand integration marketing senior people in furniture industry;
He was a visiting professor at the Beijing university of technology institute of space design.

本案从城市特征与企业发展历程的角度分析，成立20年的黎明家具公司深受着北京的数字化、信息化、现代化发展节奏和魅力的熏陶，因此为其定义一个全新概念——"数字黎明"，为其塑造出一个现代、动感、充满活力、数字化的形象，打造成一个具国际化高端的集生产、办公、展示为一体的家具中心。本案设计灵感源于企业的名字"黎明"，"黎明"也是黑暗到天明的自然景象，因此取自然景象中黑色、白色、灰色、红色作为设计的主色调，用像素点演变成黑、白、灰、红的色块作为基本设计元素，并利用无数像素点组成黎明办公英文logo"LMFU"。像素点象征信息化、数字化时代的理念，将其运用于整个空间设计中，意在创造灵动、充满活力的视觉空间，传达"数字黎明"全新概念。

This design, focusing on the urban characteristics and development history of the enterprise, that is, the 20-year-old Liming Office Furniture Company is profoundly affected by the digital, informationized and modernized development rhythm and charms of Beijing, gains a brand new concept——Digital Daybreak, aiming to deliver a modern, dynamic, vigorous and digital image of Daybreak Furniture and build an internationally top-grade furniture center integrating production, office work and exhibition. Inspired by Daybreak, the name of the enterprise, also the natural scenery from darkness to daybreak, the design takes natural black, white, grey and red as dominant hues and the black, white, grey and red color lumps developed from pixel points as basic design elements. Numerous pixel points are used and form the English logo "LMFU" of Liming Office Furniture. Pixel points imply the informationized and digital era and the application of pixel points in the whole space design is to create a dynamic visual space full of life to deliver the brand new concept of "Digital daybreak".

赎回：前罪犯的过渡中心
RE-DEEM: Transitional Hub for Former Prisoner

公司名称：马来西亚万达学院 /Organization Name : KBU International College
设 计 师：蔡嘉蕙 /Designer : Chua Chia Hui

IAI 设计之星奖

蔡嘉蕙　Chia Hui Chua
马来西亚　Malaysian

蔡嘉蕙是一名室内设计师，刚刚毕业于马来西亚万达学院（同时拥有英国诺丁汉特伦特大学的特许学位）。他认为在空间设计当中，空间现象学对人类身体的中心空间设计有很大影响。在美国的短暂游学经历当中，他发现了旅行、探索和激情对于长期的学习和个人成长起到至关重要的因素。

Chia Hui Chua is an Interior Architect who had just graduated from KBU International College (franchised degree programs from Nottingham Trent University, UK) in Malaysia. Personally and sincerely believes that phenomenology of a space is highly influential to the tenant of the space for human body is the center of a space design. Through a short journey to America, she discovered that travelling, exploration and passion are important factors for continuous learning and personal growth.

当今是什么让我们的社会变得如此的冷漠？我们怎样才能避免自己成为一名受害者？基于研究的基础上，发现部分罪犯被释放后还会再次犯罪。实施犯罪的因素很多，比如缺乏工作机会，被家庭成员所抛弃，社会歧视或者与现实生活的格格不入，可能会使前科犯继续参与犯罪。因此，对于他们来说，最大的支持是帮助他们度过被释放后的生活，是减少前科犯再犯罪的可能性的关键。救助中心在转变前科犯的未来生活中扮演着重要的角色。它为前科犯提供了一个临时避难所，临时工作机会（如自卫教练），而最重要的是一个重新连接到当今社会的生活方式、对当今社会的理解和生活知识的中心。

What made society today becoming so indifferent nowadays? And how to avoid yourself from being a victim? Based on researches, criminals are partially made up from ex-convicts that recidivate after being released. Multiple factors such as lack of job opportunity, forsaken by family members, discriminated by society and disconnected from the present may involved in recidivism of an ex-convicts. Therefore, a huge support for them to go through their life after being release is the key point to reduce the likeliness of recidivate of a former convict. RE-DEEM plays a major role in transitioning the life of ex-prisoners. It provides former convicts a temporary shelter, temporary job opportunity (as self-defense instructor), and most importantly a center to reconnect to the present society in terms of lifestyle, understanding and knowledge.

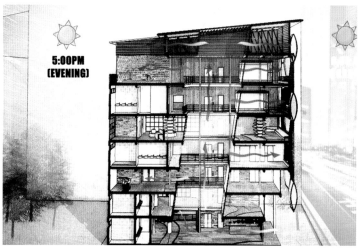

5:00PM
(EVENING)

空气呼吸中心
Air Breathing Centre

公司名称：马来西亚万达学院 /Organization Name : KBU International College
设 计 师：黄钧廷 /Designer : Wong Kwan Theng

IAI 设计之星奖

黄钧廷　Wong Kwan Theng
马来西亚　Malaysia

于黄钧廷而言，设计是一种工具，它从目标和期望出发，并且围绕着这两点来创造出新的更好的现实。他一直寻求永恒的事物。他对短期的设计或者创造一个回应潮流的设计不感兴趣。他醉心于运用现代工艺与技术来创造一个能永恒流传的室内设计，并且一直努力得到一个改变建筑内部要求，使它更加完美的灵感。

ToWong Kwan Theng, design is the means. the starting point are goals &expectations, but it surround them to create new and better reality. He always seeks for items that build-to-last. He is not a fan of short-term design or created as a response to passing trebs. He is very into modern technologies and techniques to produce timeless interior and always try to gain a sense of the changes a building's interior require for the better.

近年来，马来西亚城市地区的空气质量越来越糟糕，快速城市化已经成为空气污染的主要因素。这可能会影响人们的健康，尤其对于患病的和有呼吸困难的人群。空气呼吸中心打算帮助人们通过治疗呼吸困难，用药和健身系统来改善健康问题。除此之外，通过研讨会课程教育当地公民保护环境。

本案适应性地重用历史建筑降低的材料成本和保护环境。6层战前建筑已重用为空气呼吸中心，位于马来西亚的吉隆坡。屋顶花园教育概念为教育人们要环保，并尝试解决城市热岛问题。

在这个项目中，"肺"被选为富有表现力的元素，应用在的建筑设计及活动中。"肺"展示了呼气和吸气的过程特点，以及人体血流供应氧气的过程。在这个想法里，"肺"过滤系统已经应用在这个老房子为参观者提供新鲜空气。建筑正面为胸腔的结构，正好将肺部过滤系统包裹，参观者进入大楼时会感叹其内部的结构。

In recent years, it is noted that the air quality in Malaysia cities area are getting worse rapid urbanization has become the main factors for air pollution. This probably will affect to our health especially people who are ill and having breathed difficulties. Air Breathing Center intention to help breath difficulties people through therapy, pharmacies and gym system. Besides that, education also been provided to educate local citizen to protect environment through seminar course.

Adaptive reuse of historic building to save the material cost and environment. 6 storey pre-war building has been reuse for Air Breathing Center which is located in main cities of Malaysia, Kuala lumpur.Roop top garden provided to educate people go GREEN to protect environment and try to solving the hot island issue in cities.

In this project, "Lungs" has been selected for my expressive elements to apply in Wong Kwan Theng's building design and also part of the activities. Lungs have shown characteristic with the process of exhalation and inhalation and the flow of blood supply oxygen to human bodies. In this idea, "Lungs" filtration system has apply in this old building to provide fresh air to the visitor. Old front would remain as a ribcage to cover the lungs (filtration system) while the futuristic internal would give "wow" effect when the visitor go into the building.

生菜屋——可持续生活实验室
Lettuce House: Sustainable Lifestyle Labdecoration

项目地点：北京市顺义区京密路同乐鑫公司厂区内 /Location : Inside Tong Xinle Company Plant, Shunyi District, Beijing Dense, Beijing
项目面积：200 平方米 /Area : 200 m²
公司名称：清华大学 /Organization Name : Tsinghua University
设 计 师：刘新、贺鼎、王蔚、胡也畅、陈蔚然、苏宇融、徐浙桐、杨旭、尤婉蓉（清华大学团队）
　　　　　/Designers : Liu Xin, He Ding, Wang Wei, Hu Yechang, Chen Weiran, Su Yurong, Xu Xitong, Yang Xu, You Wanrong(Tsinghua Team)

IAI 特别奖

2014 年初，刘新老师带领清华大学美术学院项目组与绿色种植专家牛健老师合作创建了"生菜屋——可持续生活实验室"，希望将绿色、健康、低碳的生活理念应用到真实的生活场景中，从而带动更多人关注、理解并参与到可持续生活的实践中。作为一间"活"的实验室，牛健老师一家居住其中，不仅照顾各种设备、蔬菜花草，也在测试设备的运行情况，计算普通家庭的能源消耗与各类垃圾回收处理的可能性，同时为共享社区的未来参与者们提供培训。该项目最终的成果绝不仅限于运用了生态技术的住宅，而是以人为主体的可持续生活方式的实践和传播。

In the beginning of 2014, Prof. Liu Xin, leading the Project Team of Academy of Arts & Design, Tsinghua University, worked with plantation specialist Mr. Niu Jian to establish the Lettuce House, a Sustainable Lifestyle Lab, for the purpose of applying green, healthy and low carbon lifestyle into real scenes and leading more people to concerning, understanding and participating in the sustainable lifestyle practice. Actually, Mr. Niu and his families live in this lab, where he not only takes good care of all the experimental devices and plants but also tests the operation conditions of testing equipment, calculates the energy consumption of ordinary families, evaluates the possibilities of recycling different kinds of wastes, and at the meantime provides training for people intended to join the community sharing. The ultimate output of this project is not only a dwelling equipped with the ecological technologies, but also the practice and spreading of the people-oriented sustainable life style.

刘新 Liu Xin
中国 China

清华大学美术学院　工业设计系副教授；
清华大学美术学院　协同创新生态设计中心主任；
清华大学艺术与科学研究中心　可持续设计研究所副所长；
DESIS 社会创新与可持续设计联盟　清华大学美术学院 lab 负责人；
LeNS-China 中国可持续设计学习网络发起人、联合协调人。

Associate professor of Academy of Arts and Design, THU;
Director of Collaborative Innovation Center of Eco-Design;
Academy of Arts and Design, THU Deputy Director of Sustainable;
Design Research Institute of Art and Science Research Center, THU;
Coordinator of DESIS Lab THU Initiator and Coordinator of LeNS-China.

贺鼎
清华大学建筑学院博士研究生；清华大学《住区》杂志专栏负责人；爬山虎工作室联合创始人；
王蔚
九三学社、清华大学博士后、建筑学博士；
胡也畅
清华大学美术学院硕士研究生。

He Ding
doctoral candidate , School of Architecture, Tsinghua University
The charge of the magazine RESIDENTIAL,Tsinghua University,the co-founder of Boston ivy studio.
Wang Wei
Jiusan Society, postdoctoral in Tsinghua university,doctor in architecture
Hu Yechang
graduate student of Tsinghua University.

可持续生活故事版
The Story Board of Sustainable Life

共享社区发展中心
清华大学艺术与科学研究中心——可持续设计研究所

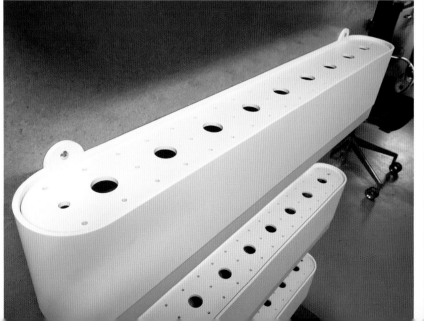

实验室系统运行图
The Lab System Operation

可持续生活实验室项目尝试从集装箱房屋建造、清洁能源利用、生活垃圾处理、中水设施与沼气系统应用、有机种植等等环节入手，构建一套综合的系统解决方案，并探索家将实验成果拓展到新型共享社区的构建中。

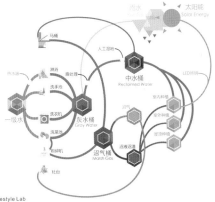

生菜屋——可持续生活实验室
Lettuce House _ Sustainable Lifestyle Lab

共享社区发展中心
清华大学艺术与科学研究中心——可持续设计研究所

半山一号别墅室内设计
BANSHANYIHAO Villa Interior Design

项目地点：江苏省无锡市滨湖区环山东路1号 /Location : No.1, Huanshan East Road, Binhu District, Wuxi, Jiangsu
项目面积：600平方米 /Area : 600 m²
公司名称：苏州工业园区九寸钉空间设计工作室 /Organization Name : Nine Inch Nails Design Studio
设 计 师：卢晓晖 /Designer : Lu Xiaohui

IAI 设计优胜奖

卢晓晖 Lu Xiaohui
中国 China

2014年 南京131工程室内设计；
2013年 江苏大剧院室内设计；苏州国际博览中心室内设计。
2011年至今 苏州工业园区九寸钉空间设计工作室（华鼎装饰设计院）设计师；
2008—2011年 西南交通大学 设计艺术学 研究生；
2004—2008年 西南交通大学 环境艺术设计 本科。

2014　131 Program,Nanjing;
2013　Jiangsu Grand Theater, Suzhou International Expo Center.
2011　Designer of Suzhou Gold Mantis Construction Decoration Co.,Ltd.;
2008—2011　Postgraduate of Department of Art,Design,Southwest Jiaotong University;
2004—2008　Undergraduate of Department of Environmental ;
Art Design, Southwest Jiaotong University.

半山一号别墅群位于无锡太湖风景区马山半岛，是全国十二个国家级旅游度假区之一。设计以"半"字为出发点，"半半哲学"是调和了儒家哲学和道家哲学的一种中庸生活，"所以理想人物，应属一半有名，一半无名"；做人如"半杯水"，主张谦虚；居住主张一半建筑，一半自然。
半山之间有栈道，栈道有曲折，这是设计的另一个隐喻之处，也是设计的灵感所在。尼采说："别在平原上停留，也不要爬到极顶。打从半山腰处看世界，世界会显得更美好。"你还会带着对山顶景色的无限憧憬与遐想，快乐地行走在途中。于是设计师把半山之间的栈道搬进了客厅，将原建筑内向的楼梯改到了客厅的中央，并错落有致地通往二楼，形成一个中部中庭景观。一层的空间中墙体消失了，而是这座"栈道"将空间分为了四部分：分别是左客厅（一半看户外）、右客厅（一半看室内）、餐厅、书房、琴室……简约至极的设计语言却表达了丰富的空间层次与精神内涵。

BANSHANYIHAO villa located in Wuxi Taihu lake scenic spot masan peninsula, villa near the taihu lake, has the very good landscape. Design with "half" word as a starting point, form a unique design concept. "Half" is the Chinese attitude toward life and the way of life, the person such as "half a glass of water", advocate modest, live like "architecture", half of the buildings, half natural.
Mid-levels between road, the road has twists and turns. This is the inspiration of design. Stylist the mid-levels between plank road into the living room, change the original building stairs to the middle of the sitting room, the space is divided into four parts were left sitting room half (outdoor), right half (indoor) sitting room, dining-room, study、practice room... Contracted so the design of the language are expressed the most abundant space level and spiritual connotation.

张家港小城河改造
Zhangjiagang Town River Reconstruction

项目地点：江苏省张家港市 /Location : Zhangjiagang, Jiangsu
项目面积：65 000 平面图 /Area : 65 000 m²
公司名称：澳大利亚·柏涛景观 /Organization Name : Botao Landscape
设 计 师：王珂 /Designer : Wang Ke

IAI 设计优胜奖

柏涛景观（总部：澳大利亚）
BOTAO Landscape (Headquater:Australia)
中国 China

澳大利亚·柏涛景观属于澳大利亚柏涛（墨尔本）建筑设计有限公司全球机构的成员，主要承担国内外各类环境景观设计项目，成立于1999年，是深圳较早一批成立的大型景观设计公司，通过了ISO9001：2008质量管理体系认证，具有风景园林工程设计专项乙级资质。由外籍规划师、景观设计师协同工作，百余位景观师、园艺师、高级建筑师精心从事项目方案到施工图的设计工作，曾获得金地集团"2013年度优秀合作伙伴"、佳兆业集团"2012年度优秀合作单位"。

BOTAO Landscape (Australia) belongs to Peddle Thorp Architects Melbourne. The company was established in 1999 and was one of the first large landscape design companies in Shenzhen. It has passed the "ISO9001: 2008 quality management system", and also has possessed the grade B qualification of landscape architecture construction design certification. BOTAO Landscape (Australia) is specialized in environmental and landscape art design. It has rich project design experience not only in China, but also in the whole world. Design covers public cityscape design, residential landscape design, urban complex design and tourism space planning. The company has also received many honorary titles, such as "The Best Collaboration 2013" from Gemdale group, "The Best Collaboration 2012" from KAISA group.

小城河位于城市核心商业区步行街的南侧，东起谷渎港、西至港城大道，全长约2200米，在这一个世纪里，水质污染严重，周边环境凌乱。

小城河综合改造工程以打造核心商业区的"城市客厅"为理念，在治污引清的同时恢复河道自然生态，在建设景观的同时提供更加宜人的环境，在综合整治的同时全面提升整个区域的基础设施功能。

立足于现代简约风格，小城河、谷渎港的改造从建筑立面的设计、石材选型到景观铺装，处处洋溢着浓郁的水乡风貌，以亭、榭和片墙元素，演绎江南特有的枕河人家。老杨舍人最熟悉的青龙桥得以修复，以青龙桥、龙吟、谷渎潮声、绿香亭、竹筏码头、轮船码头、暨阳门城墙及"八不准"碑等老杨舍历史上原有的八大文化元素，通过创意设置，点缀在谷渎港河道两侧，展现谷渎港滨水人文景观特色。

整个景观、建筑设计，摒弃一切烦琐的符号，留下最安静的价值感。

Town river is located in the south of commercial pedestrian street in the town center. From the Gang Du Harbor to Harbor City Street,it has been 2000milis ,but it has been facing the serious water pollution and the environment pollution for a century.
The comprehensive renovation project of Town River is build around the concept of urban living room ,handle the pollution and bring in clean water as well as recover the natural ecology.In the construction of the landscape ,it will provide a more pleasant environment.At the comprehensive environment improvement,the basic infrastructure capabilities in entire region will be improved.
Based on the Modern Concise Style,the reconstruction on Gang Du Harbor in Town River play the kind of style that people who live nearby river in Jiang. From the design of building elevation to the sculpt of stone and the aesthetics pavement,everywhere is brimming with typical water village style with the element of kiosk,shed and wall.The most familiar"Qing Long Bridge"has been repaired.And the original eight culture elements including QingLong Bridge,Dragon Voice,Gu Du Chao Shen,Lu Xiang Kiosk,Bamboo Raft wharf ,Ji Yang Men City Wall and the"Ba Bu Zhun"stele,show up along the two side of Gang Du Harbor by the creative settings.It shows the cultural landscape features of GangDu Harbor.
All the landscape,architecture design without the complete symbols,just leave the most quiet sense of worth.

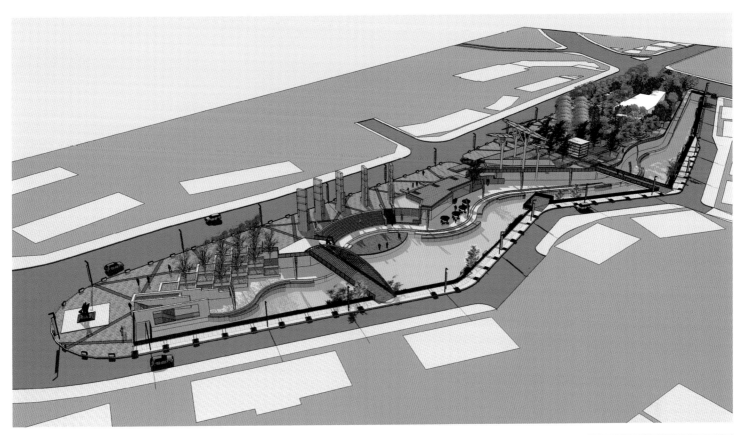

财富天地商业空间软装
Fortune World Commercial Space Decoration

项目地点：广东省广州市西湾路 150 号 /Location : 150 Xiwan Road, Guangzhou, Guangdong
项目面积：260 000 平方米 /Area : 260 000 m²
公司名称：广州大学纺织服装学院 /Organization Name : College of Textile and Garment, Guangzhou University
设 计 师：乔国玲 /Designer : Qiao Guoling

IAI 设计优胜奖

乔国玲　Qiao Guoling
中国　China

广东省顺德美的集团 设计师；
广州大学纺织服装学院主任；
兼任广州市湛艺装饰工程有限公司总设计师；
长期从事陈设计和中外艺术设计交流的教学与研究；
率先在中国高等院校开设"室内陈设设计"课程；
出版《室内陈设艺术设计》和发表学术刊物论文多篇；
作品入选《中国环境设计年鉴》- 为中国而设计。

Designer in GuangDong ShunDe Midea Group;
Director of the GuangZhou Textile Garment Institute;
Chief Designer in Guangzhou Cham Art Decoration Engineering co., LTD;
Long engaged in display design and art design teaching and research of the communication;
The first indoor display design courses in colleges and universities in China;
Published Indoor display art design, and many academic journal papers published articles;
Works to be included in the China environmental design yearbook - designed for China.

项目是广州市老城区广州市水泥厂的改造升级，由香港上市公司越秀地产投资数十亿元来打造成广州市城西最大的主题商城财富天地广场。项目初始，甲方希望我们通过各种方式来烘托这个 26 万平方米室内的商业气氛。设计师通过对建筑形态、行业业态和商业空间的动线分析提出设计目标：打造一个具有主题性的、时尚的、创造性的、文化性的商务综合体。设计提炼了其他设计师对布艺的认识来装饰这个大型的商业空间，最终用最少的经费达到了大家比较满意的装饰效果。

The project is the old city of Guangzhou Shi Min Shi factory (Guangzhou cement) upgrade, by the Hong Kong listed companies of Yuexiu real estate investment Shujishiyilai into Guangzhou Chengxi biggest theme mall fortune world plaza. Initial project, Party A shall designers hope that through various ways to foil the 260 000 square meter indoor commercial atmosphere. Designers passed on the architectural form, industry and commercial space line analysis proposed on the basis of our design goals: business complex culture to create a theme, with the fashion, creative, the. Refining the other designers understanding of cloth to decorate the large commercial space, finally with the least amount of funds reached satisfactory decoration effect.

合作伙伴 Cooperation Partners

支持机构 Media Supporter

图书在版编目（CIP）数据

2014—2015 IAI 设计奖年鉴 / 何昌成 编. —武汉：华中科技大学出版社，2016.1
ISBN 978-7-5680-1450-2

Ⅰ.① 2… Ⅱ.①何… Ⅲ.①室内装饰设计－作品集－中国－现代 Ⅳ.① TU238

中国版本图书馆 CIP 数据核字 (2015) 第 300493 号

2014-2015 IAI 设计奖年鉴

何昌成 编

出版发行：华中科技大学出版社（中国·武汉）	
地　　址：武汉市武昌珞喻路 1037 号（邮编：430074）	
出 版 人：阮海洪	
责任编辑：曾　晟	责任监印：秦　英
责任校对：徐　茜	美术编辑：易凌燕

印　　刷：天津市光明印务有限公司
开　　本：965 mm×1270 mm 1/16
印　　张：20
字　　数：288 千字
版　　次：2016 年 1 月第 1 版第 1 次印刷
定　　价：338.00 元 (USD 69.99)

投稿热线：(010)64155588-8000
本书若有印装质量问题，请向出版社营销中心调换
全国免费服务热线：400-6679-118 竭诚为您服务
版权所有　侵权必究